A History of the Transport Service

Adventures and Experiences of United States Transports and Cruisers, in World War One

By Vice Admiral Albert Gleaves

Commander of Convoy Operations in the Atlantic 1917-1919

PANTIANOS
CLASSICS

Published by Pantianos Classics

ISBN-13: 978-1-78987-076-3

First published in 1921

Contents

THIS BOOK IS DEDICATED TO THE OFFICERS AND MEN OF THE CRUISER AND TRANSPORT FORCE WHO ACHIEVED WHAT THE ENEMY THOUGHT IMPOSSIBLE.

Preface

There seems no excuse for offering the public another book of personal Memoirs of the Great War; but so much has been written about all the different phases of preparation and action, and so little of the actual transportation of the troops, which made the fighting possible, that I have yielded to the persuasion of friends and shipmates to add my contribution to the daily increasing stories of the events of 1917-18. I do this because in a measure I may be able to show my appreciation of the hard work done by the officers and men of the American Transports, and my admiration for their unsurpassed skill and endeavor in the performance of their duties. At sea almost constantly, in the severest weather that has swept the Atlantic Ocean for many years, these Master Mariners of the United States lived up to the highest traditions of the sea, and brought credit to their country. For the most part this story is told by them, in their own words, and so far as possible taken from their official reports.

In thus presenting the narrative the book will have served its purpose if it throws a light on the character and professional ability of those officers and men of the Navy who had the good fortune to take our gallant Army to France.

I am much indebted to Commander C. C. Gill, U. S. N., my Aide and Flag Secretary, for his wise counsel, his unfaltering assistance while preparing the manuscript, and for his literary skill in smoothing out my patchwork in getting into shape a mass of material which sudden orders to sea forced me to leave confused and unfinished.

Albert Gleaves,
Admiral U. S. Navy.
Commander in Chief,
United States Asiatic Fleet.

U. S. Flagship *South Dakota,*
 Vladivostok, Siberia,
  13th January, 1920.

Introduction - The Crisis of 1917

In April, 1917, at the time of the war declaration of the United States, the Allied cause was in serious danger. Apparently Germany had victory within her grasp. Both on land and at sea William Hohenzollern was at the zenith of his power. France was on the verge of collapse. Great Britain, dazed by the submarine blow struck at her trade and shipping, found her sea supremacy challenged and the great British Navy unable to protect fully the commerce essential to England's existence.

Had the German genius been equal to the role, this would have been the year for the supreme effort of Prussian Militarism. But the German General Staff was contemptuous of the unprepared, peace-loving people across the Atlantic. The War Lords miscalculated the spirit and fighting abilities of the American people. They had applied their own formulas in reckoning with a nation totally unlike their own. This was their fatal error. When Ludendorff in the Spring of 1918 launched his great offensive, it was too late. By that time the U-boat had been checked and Allied supremacy of the sea reestablished. This marked the wane of Prussian power.

The fierce attacks and temporary successes of the enemy on the Western front in the Spring of 1918 represented the final desperate effort to wrench victory from defeat. It was doomed to failure. The weight of America's potential power was already beginning to tip the scales. A great army had arisen; it was being spirited across the seas, and a few months later, at the Marne, it met, turned, and routed the best troops of Germany. This reverse shook the Hohenzollem throne, and served notice to all the world that the strength of the United States Army was equal to its task.

It was not only the effective fighting of our Army which contributed so much to win the war. Even more overwhelming was the surprise of its presence, its demonstrated ability to fight, and the conviction forced upon the German command that there was an unending stream of the same fighting power pouring in upon the battlefields.

All this, however, came one year after the crisis of 1917. Judgment in retrospect is often influenced by the light of later events. In view of what has happened since May, 1917, to the casual inquirer it now appears to have been a natural and obvious course, that the United States should have devoted all its resources to raising, equipping and transporting a great army. Analysis and reflection, however, show that this was not an ordinary enterprise either in conception or in execution. On the contrary, it was unique and remarkable. There is little risk of hyperbole in venturing the opinion that the raising, transporting and supplying overseas of this army of two million men will be finally ranked as one of the greatest achievements in the annals of history.

Turning back now to the Spring of 1917, we find that on land German arms were ascendant on all fronts. In the East, Russia had been almost eliminated as a military factor; Roumania, Serbia and Montenegro had been conquered; Bulgaria and Turkey, although ostensibly Allies of Germany, were actually under the Prussian yoke; the aims of the German Eastern Policy, which included a German Mittel-Europa, had been realized, and it only remained to compel the Western countries to recognize them. In the West, although foiled in the attempt to crush France, German arms had attained considerable success; Belgium and Luxemburg had been overrun; Holland had been isolated; and a valuable strip of Northern France had been occupied.

On the Western front, likewise, the adverse factors in the military situation must be appreciated; it is true that Joffre had stopped the enemy at the Mame in 1914, and that the German offensive against Verdun in 1916 had failed, but, on the other hand, the much-heralded Allied offensive for the Spring of 1917 was at that time also ending in failure. The hope of Allied victory aroused by German readjustments along the Hindenburg line had been quenched by their stubborn defense in the Battle of Arras (April-May, 1917), and it had become evident that success in "breaking through" the German positions was no nearer than it had been before the costly Battle of the Somme.

The Italians had also been unable to develop a successful offensive. In fact, both in the East and in the West the Central Powers were showing ability to hold fast all the great military advantages they had gained. It is not surprising that enemy hopes ran high, while the Allied peoples were depressed.

Nor did the enemy fail to exploit their advantage. For long years they had prepared for this opportunity and the German Government had a special army of secret agents and political hirelings scattered throughout foreign countries instructed to kindle sedition and undermine Allied morale. In this connection, it is interesting to recall the *"Spurlos Versenkt"* [1] incident, the Zimmerman note (scheming the return of New Mexico, Arizona, California, and Texas to Mexico), the rumor of a German-Japanese treaty, and, most significant of all, the political disintegration — almost always a consequence of military disaster in the field — proceeding in France in the Spring of 1917. Even now, few people in this country appreciate that enemy agents had attained such power in Paris that they then worked almost in the open, spreading corruption both in the homes and in the trenches. When French regiments mutinied and the Commanders in the field reported explicitly to their government the sources from which sedition was propagated, officials either would not or dared not take prompt effective action. All this was revealed at the trial of Malvy, then Minister of the Interior, and charged with safeguarding France against enemy machinations. He was finally convicted for neglect of duty and banished. Also, the then Head of the Secret Police, together with the Assistant Prefect of Police in Paris, were later sentenced to prison for intercourse with the enemy during this period. It was not until after the first American Expedition had landed in France and Clemenceau had been swept into power that these alarming conditions were fully exposed and dealt with effectively.

Bad as was this 1917 situation on land, the situation on the sea was even more threatening. The German Government had broken away from international law and on February 1, 1917, had begun a campaign of unrestricted submarine warfare. This had been planned as the great German offensive of 1917. It was a blow aimed at the vitals of the Allies, their lines of communication, and so careful and thorough had been its preparation that it came perilously near to success. When the United States entered the war German U-boats were sinking merchant ships at the alarming rate of 800,000 tons a month. This placed England in desperate straits and unless these submarine depredations could be checked, the time that Great Britain could hold out was simply a matter of mathematics. Nor was this the only success scored by the German Navy. The more or less prevalent conception that German Naval Power took no important part in the war is erroneous. The cover of German dreadnought guns permitted the U-boats to slip out on their deadly purpose: the High Seas Fleet had accomplished an essential mission in preventing a close blockade, in converting the Baltic practically into a German Lake, in lending aid to the Armies advancing against Russia, and in keeping the great British naval establishment busy in the North Sea.

Also of far-reaching consequence had been the success of the *Goeben* and *Breslau* in escaping to Constantinople. This influenced Turkey to enter the war and contributed to the defeat suffered by the Allies in their attempt to force the Dardanelles. Thus did the German Navy blockade Western and Southern Russia. Communications via the Northern and Eastern ports of Russia were slow, inadequate and uncertain.

Such in brief were conditions on land and on sea when the United States entered the war. The leaders of France and England were keenly alive to the crisis existing at this time and the need of impressing it upon the government of their new associate. Both countries lost no time in dispatching military and naval envoys to Washington. In April, 1917, the Joffre Mission arrived from France and the Balfour Mission from England. Their object was to urge the prompt and active assistance of the United States.   The Allied Admirals requested the Navy Department for as many destroyers as possible, but said "even one" would suffice to serve notice to the Central Powers that the United States was in the war in earnest. With the same desire for moral effect, Joffre pleaded for at least one division of the Army to be sent to France at once.

The first fruits of these Allied appeals were the dispatch of six destroyers to Europe. The readiness of these destroyers for foreign service, when the call came, was a matter of gratification to the entire country. They were followed at intervals by nearly the entire destroyer force, only a few of the older type being retained for patrol and escort duty on our own coast.

No doubt Germany anticipated that her unrestricted submarine offensive would cause the United States to enter the war. This was not, however, a sufficient deterrent. Germany was confident that her U-boats would prevent the United States from taking an effective part. The enemy counted upon our unpreparedness and did not think it possible for this country to transport and supply overseas a large army. And it should be added that some of the leading Allied strategists took a view hardly more sanguine as to the extent of America's war effort. [2] In the glamour of our overwhelming victory there is a tendency in present-day discussion to underrate the product of the deliberations of the expert German General Staff which directed the enemy policy as well as military effort. It must be admitted by unprejudiced opinion, based on a measured view of the military situation in the Spring of 1917, that Germany had at least plausible grounds for her conclusions. (The German point of view is given in detail in a German Admiralty memorandum reproduced in the appendix.)

The German Staff argued that in the face of their submarine campaign and the consequent shortage of shipping which was already causing embarrassment to England and France, the United States would never venture the overseas transportation and supply of an effective army; and if they did venture this hazardous undertaking the U-boats would see to it that disaster should attend and make impossible its accomplishment. Few informed critics will venture to attack this opinion as altogether illogical. On the contrary, the experience of previous wars viewed in the light of the then current official utterances and state papers, seemed rather to support the argument that the interest of the United States in the issue did not constitute a sufficiently impelling motive to induce the government to make a decision involving a great risk.

The decision to send the United States Army to the Western front was made, however, and ultimate victory or defeat hinged upon whether or not this decision — the overseas transportation and supply of an effective army — could be successfully carried out.

[1] *"Spurlos Versenkt,"* the conception of German diplomacy by which all evidence of U-boat unscrupulousness was to be obliterated by sinking without leaving a trace of ship, crew, or passengers.

[2] On 19th July, 1017, the British Naval Authority Admiral Boresford expressed himself as follows: "At the present rate of losses — British, Allied, and Neutral, average from 1st of February to 14th July (say about six months) — I make out that British, Allies and Neutrals are losing ships at the rate of about seven million tons a year. I also make out that if the allied shipbuilding firms of the world put forward their full strength as at present, they could not produce more than four million tons of now shipping, in other words about one-half.   I am also distressed at the fact that it appears to me to be impossible to provide enough ships to bring the American Army over in hundreds of thousands to France, and, after they are brought over, to supply the enormous amount of shipping which will be required to keep them full op with munition, food and equipment."

Chapter One - The Naval Mission— Summary of Transport Operations

The work of the Navy in connection with the transportation of troops to France constitutes a distinctive feature of the World War. As has been pointed out in the Introduction the attending political and military circumstances incident to the collapse of Russia, the critical situation on the Western front, and the threat of the German submarine combined to make the undertaking of special significance, and throughout the year following the entry of the United States into the war the military and naval developments were such that the safe transportation across the Atlantic of troops and supplies became a problem of more and more pressing importance.

The United States Army in France was a decisive factor in obtaining speedy victory. The transportation of this army overseas under naval protection was therefore, a major operation of first importance. A large share of this urgent mission devolved on the United States Navy, and its successful accomplishment in the face of great difficulties is another page to the record of the service in keeping with its past history and traditions.

Much confusion of thought has existed as to just how the vast work of transporting a United States Army numbering 2,079,880 souls to Europe was accomplished. It is unfortunate that misinformation should be disseminated respecting an operation in which the different organizations concerned performed their respective functions in utmost harmony and cooperation. All did their allotted parts splendidly and efficiently. All share in the satisfaction resulting from the successful accomplishment of a difficult and urgent undertaking.

Inasmuch as the principal field of British naval activities was the North Sea and English Channel — the Grand Fleet containing the enemy High Seas Fleet was an essential condition without which neither troop transports nor freighters could have sailed — the task of breaking the U-boat blockade in the Atlantic naturally became the immediate mission of the United States Navy. The prompt dispatching of destroyers, coast guard cutters, yachts, and all other available craft of a type useful against the submarine, to the East Atlantic, and the splendid work accomplished by these vessels and others later sent to augment their strength, in cleaning up these waters of U-boat devastation is a matter of record, the importance of which in winning the war is conceded from all quarters. This was the first step in preparation for sending the United States Army overseas.

The next step was the development of the transport service and the convoy and escort system. In this work the Cruiser and Transport Force cooperated with the destroyers and other anti-submarine craft abroad. In addition.

Great Britain, France and Italy supplied troop ships. As would be expected from Great Britain's enormous merchant marine, she was able to supply the greatest carrying capacity. She had the ships ready for this use, and 48 ¼ per cent of the American Army were transported in British steamers; 2 ½ per cent were carried in French ships, and 3 per cent in British leased Italian vessels. The remaining 46 ¼ per cent were carried in United States ships, and all but 2 ½ per cent of these sailed in United States naval transports.

All the troops carried in United States ships were escorted by United States men-of-war; that is, cruisers, destroyers, converted yachts, and other anti-submarine craft. Also for the most part the troops carried in British, French and Italian ships were given safe conduct through the danger zones by United States destroyers. Roughly, 82 ¾ per cent of the maximum strength of the naval escort provided incident to the transportation of United States troops across the Atlantic was supplied by the United States Navy, 14 1/8 per cent by the British Navy, and 3 1/8 per cent by the French Navy.

In making our Army in France effective, special mention should be made of the Naval Overseas Transportation Service. Little could have been accomplished without these unromantic, rusty, slow plodding tramps, transporting food, munitions and supplies.

It was one of the anomalies of this war due to our small deep sea merchant marine that instead of merchant sailors being called upon to help man our battleships, the war fleet was called upon to help supply trained officers and men for transports and freighters.

The newly-created Emergency Fleet Corporation and the Shipping Board did their best, and indeed accomplished wonders in quickly building and equipping a vast merchant fleet; but in obtaining civilian crews to operate the ships they were heavily handicapped by labor conditions and the lack of trained seamen. When the shipping board turned to the Navy for assistance, that our naval establishment was ready to operate successfully an enormous merchant fleet was a war-winning factor which is now too frequently overlooked.

In the latter part of 1917, the Naval Overseas Transportation Service was organized under the Navy Department and supervising branch offices were established in the principal Atlantic ports; 72 vessels were originally assigned to this service but when the Armistice was signed the Navy had a cargo fleet of 453 ships, including 106 ready to be taken over. The Naval Personnel operating these ships numbered 5,000 officers and 45,000 enlisted men. This cargo fleet was additional to the fleet of troop transports also manned by the Navy.

The crews of some of these freighters endured the greatest hardships of the war. They faced not only the menace of the U-boat, but also the perils of the deep, danger of collision and shipwreck, and the liability of instant death by accident from inflammable and explosive cargoes.

Nor should the Ship Control Committee be forgotten. This Committee, headed by Mr. P. A. S. Franklin, was charged with loading the cargo and quick turn-around of the vessels. The services of Mr. Franklin and his assistants are part of the history of war transportation and supply. To quote the expression of General Shanks (Commanding General at the Port of Embarkation in New York), "Our great embarkation service was of a threefold character, the Navy, the Army, and the Ship Control Committee."

At the time of our entry into the war, although we had a very small deep-sea Merchant Marine, our coastwise and inland shipping industry was a flourishing one. The ships for the most part were unsuitable for overseas work, but the personnel trained in them was of excellent quality and proved an invaluable accession to naval strength. I think I can speak with authority of the fine work done by both officers and men. In the Cruiser and Transport Force the Merchant Marine and the Navy worked together, without difference or distinction, shoulder to shoulder, "all of us together on the capstan bars."

The outstanding lesson which the experience of the war has driven home is the value both in peace and in war of a prosperous deep sea Merchant Marine.

SUMMARY OF TRANSPORT OPERATIONS

Previous to 1917 the idea of a United States overseas expeditionary force numbered by millions was regarded as a remote if not impossible contingency. Consequently no extensive peace-time preparations had been made for such an undertaking. The declaration of war with Germany found the United States without a transport fleet and without a merchant marine capable of supplying ships for transporting a large military expedition. At this time the Cruiser and Transport Force had not been developed. Of the twenty-four cruisers later gathered together for war service in this command, some were in the Atlantic Fleet, some on special duty, some unassigned, while still others were in the navy yards in reserve, manned by reduced complements. Out of the forty-eight naval transports engaged in carrying troops abroad only two were then in the naval service. These were the *Henderson,* still under construction, but nearing completion, and the *Hancock,* an old vessel of slow speed, later withdrawn from overseas transport duty because of her unseaworthy condition. The Army had a few transports but they were not suitable or ready for transatlantic service.

Thus it was that our Navy entered a new field of operations. Without warning, the Navy Department and the War Department were confronted with the problem of sending to Europe hundreds of thousands of soldiers — how many, no one could tell. Joffre, in an interview with the Secretary of War in May, 1917, said that he thought that 400,000 would be our limit, and that one French port would be sufficient to receive them. How amazed he would have been could he have looked into the crystal and seen what this country transported to France in men and material during the next eighteen months.

It is a remarkable and noteworthy example of American ingenuity and zeal that, starting with almost nothing at the beginning of the war, a United States naval transport service was built up which carried almost a million soldiers to Europe. In spite of the determined efforts of submarines to prevent it, their numerous attempts were frustrated and these troops were landed in Europe. This was accomplished without the loss of a single soldier by the hand of the enemy. [1]

The splendid cooperation of the Army made this possible. The Army organized and developed an efficient system for loading and unloading the ships at the terminal points. The Navy transported the troops and safeguarded them en route.

General Ludendorff's book reveals the promises made by the German naval chiefs that their submarines would prevent the transportation of United States troops. When the soldiers began to arrive the German people and the German press began to complain bitterly that these promises had not been kept. In spite of renewed promises and redoubled efforts our transports continued to arrive without losses while East-bound. This resulted in the fall of von Capelle. His successor, Admiral von Mann, was equally unsuccessful, although for a time submarines were diverted from trade routes to concentrate their efforts against our transports. When the British steamer *Justicia* was sunk the German Admiralty officially announced that it was the *Leviathan* loaded with American soldiers. The consequent jubilation in the German press, followed by depression when the truth was ascertained, is an indication of how eagerly the enemy sought to revive the morale of the people by successful operations against our troopships.

On homeward-bound voyages, however, the Navy was not so fortunate. In a measure this was due to need of concentrating maximum naval escort protection on troop-laden convoys. Frequently this necessitated lighter escort for the ships returning, and it was on these homeward-bound vessels that the submarines scored their successes. The United States Transports *Antilles, President Lincoln,* and *Covington* were torpedoed and sunk. The *Finland* and *Mount Vernon* were torpedoed, but were able to reach port for repairs. The United States armored cruiser *San Diego* struck a mine laid by a German submarine and was sunk.

The service was not without hazard, as is shown by the fact that more than half of the war casualties in the United States Navy were suffered in the Cruiser and Transport Force. This was to be expected because the large deep draft ships were the chief prey sought by the enemy U-boats. Nor were the enemy guns and torpedoes the only menace — danger from fire and internal damage was enhanced by the machinations of enemy secret agents, and the likelihood of collision was increased by the necessity of maneuvering without lights in convoy formation vessels manned for the most part by inexperienced crews.

15

On November 11, 1918, when the Armistice was signed, the Cruiser and Transport Force of the United States Fleet numbered twenty-four cruisers and forty-two transports, manned, exclusive of troops carried, by about 3,000 officers and 42,000 men. This is in addition to the 453 cargo ships which the Navy manned and operated with 5,000 officers and 45,000 enlisted men.

After the signing of the Armistice the United States Transport Fleet expanded still more, and developed into a fleet of 149 ships manned by 4,238 officers and 59,030 men, with the gratifying result that 86.7 per cent of our overseas army was brought home under the Stars and Stripes.

The return movement began immediately after the Armistice and continued at a much more rapid rate than was attained in going over. In June, 1919, our Naval transports, which had been increased in number by seventy-one ships, brought back in 115 ships 314,167 combatant troops, while foreign ships carried an additional 26,825. The maximum number transported during the war — by all ships in one month — was 311,359. From November, 1918, to July, 1919, a total of 1,493,626 had been returned to the United States. The older battleships and armored cruisers were also used in the repatriation of our soldiers. The United States Navy alone transported across the Atlantic a grand total of 2,511,047 soldiers.

The scope of this book is a brief narrative of the adventures and achievements of the United States Cruiser and Transport Force compiled from the official files with such explanatory notes and summaries as appear necessary for reason of coherence and clearness.

It has been authoritatively said that the best way to write history is to let those who made it tell their own story in their own words. In the following pages it has been attempted to use this method and they are made up largely of quotations from those who took part in the operations.

[1] This refers to the U. S. Naval Transport Force. U. S. soldiers embarked in the British steamer *Tuscania* were lost when that vessel was torpedoed and sunk and 102 more were killed or drowned when the American freighter *Ticonderoga* was sunk by a U-boat.

Chapter Two - The First Expedition, Preparation

At the time the United States entered the war I was in command of the Destroyer Force of the United States Atlantic Fleet and had had this assignment for about one year and a half. I was summoned to Washington on May 23, 1917, and informed by the Secretary of the Navy that in addition to my other duties I had been selected to command the first expedition to France. On May 29th, I received formal orders designating me "Commander of Convoy Operations in the Atlantic."

My flag was then flying in the armored cruiser *Seattle,* and I proceeded immediately to New York to expedite preparations. The 9th of June had been set for the sailing date, but, after a personal inspection of the ships which the War Department had chartered and was converting into transports, and having conferred with the Army Quartermaster, I recommended that the sailing date be advanced to June 14th; this date was selected not without consideration of the phase of the moon as affecting night submarine attack at the expected time of arrival off the French Coast.

The Navy then had three vessels available for troop transport work, the *Hancock,* the *Henderson* and the recently seized German converted steamer *Prinz Eitel Friedrich,* renamed the *DeKalb,* The Army had a few regular transports, but none were suitable and ready for trans-Atlantic convoy operations. It was necessary to commandeer such ocean-going vessels as could be found and alter them as quickly as possible for carrying troops.

Unfortunately we had no adequate deep-sea Merchant Marine to draw upon and the somewhat motley assemblage of ships finally gathered together for the first expedition did not long survive the duty imposed upon them. Some were torpedoed, others relegated to carry cargo and cattle, and some were subsequently wrecked or dropped out altogether because of unseaworthiness.

Looking back to the first expedition of June, 1917, it seems indeed that the hand of Providence must have been held over these "arks" or the task never could have been accomplished. Who would have dreamed at that time that we were laying the foundation of the greatest transport fleet in history?

As the terms "train," "escort," and "convoy" are somewhat confusing to those unfamiliar with naval terminology, it might be well to define them. "Train" refers to a body of troopships or cargo ships or other vessels requiring protection and making passage in company. The term "escort" designates the fighting ships which accompany and protect the "train." The entire assemblage of ships consisting of both "train" and "escort" comprise a "convoy." For example, we would speak of a "convoy" of twelve ships including the "train" of six transports with an "escort" of one cruiser and five destroyers, or "escorted" by one cruiser and five destroyers.

The first expedition comprised four convoys made up as follows:

CONVOY GROUP I

TRAIN	ESCORT
Troopship Saratoga	Armored Cruiser Seattle.
(Lt. Comdr. L. Coxe)	(Captain D. W. Blamer)
" Havana.	Auxiliary Cruiser De Kalb.
(Comdr. J. R. Defrees)	(Comdr. W. R. Gherardi)
" Tenadores.	Converted Yacht Corsair.
(Comdr. W. R. Sexton)	(Comdr. T. A. Kittenger)
" Pastores.	Destroyer Wilkes.
(Comdr. B. B. Bierer)	(Lt. Comdr. J. C. Fremont)
	Destroyer Terry.
	(Lt. Comdr. J. F. Shafroth, Jr.)
	Destroyer Roe.
	(Lt. Comdr. G. C. Barnes)

17

CONVOY GROUP II

TRAIN	ESCORT
Troopship Momus.	Scout Cruiser Birmingham.
(Comdr. W. N. Jeffers)	(Comdr. C. L. Hussey)
" Antilles.	Converted Yacht Aphrodite.
(Lt. Comdr. D. T. Ghent)	(Lt. Comdr. R. P. Craft)
	Destroyer Fanning.
" Lenape.	(Lt. A. S. Carpendar)
(Lt. Comdr. P. E. Dampman)	Destroyer Burrows.
	(Lt. H. V. McKittrick)
	Destroyer Lamson.
	(Lt. W. R. Purnell)

CONVOY GROUP III

TRAIN	ESCORT
Troopship Mallory	Cruiser Charleston.
(Lt. Comdr. G. P. Chase)	(Comdr. E. H. Campbell)
	Armed Collier Cyclops.
" Finland.	(Lt. Comdr. George Worley)
(Comdr. S. V. Graham)	Destroyer Allen.
" San Jacinto.	(Lt. Comdr. S. W. Bryant)
(Lt. Comdr. S. L. H. Hazard)	Destroyer McCall.
	(Lt. Comdr. L. M. Stewart)
	Destroyer Preston.
	(Lt. j.g. C. W. Magruder)

CONVOY GROUP IV

TRAIN	ESCORT
Cargo Ship Montanan	Cruiser St. Louis.
(Comdr. P. N. Olmstead)	(Comdr. M. E. Trench)
	Cruiser Transport Hancock.
" Dakotan.	(Captain E. T. Pollock)
(Comdr. C. Shackford)	Armed Collier Kanawha.
" El Occidente.	(Lt. Comdr. R. Werner)
(Lt. Comdr. H. W. Osterhaus)	Destroyer Shaw.
	(Lt. Comdr. M. S. Davis)
" E. Luckenbach.	Destroyer Ammen.
(Lt. Comdr. A. C. Pickens)	(Lt. G. C. Logan)
	Destroyer Flusser.
	(Lt. j.g. R. L. Walling)
	Destroyer Parker.
	(Lt. Comdr. H. Powell)

The table following shows the pre-war owners and employment of vessels used in this convoy:

Name	Gross Tonnage	Line	Duty
Saratoga.........	6391	New York & Cuba Mail S. S. Co.	Mail Steamer
Havana..........	6391	New York & Cuba Mail S. S. Co.	Mail Steamer
Tenadores........	7782	Tenadores S. S. Co.	United Fruit Co. Line
Pastores.........	7781	Pastores S. S. Corp.	United Fruit Co. Line
Momus..........	6878	Southern Pacific Company	....
Antilles..........	6878	" "	....
Lenape..........	5179	Clyde S. S. Co.	....
Mallory..........	6063	Mallory S. S. Co.	....
Finland..........	12,229	Dampfsch. Ges. Argo.	....
San Jacinto.......	6069	Mallory S. S. Co.	....
Montanan........	6659	American S. S. Co.	Cargo carrier
Dakotan.........	6657	American & Hawaiian S. S. Co.	....
E. Luckenbach...	2730	Luckenbach S. S. Co.	Cargo carrier

The arrangement of bulkheads and compartments in most of these transports was not at all satisfactory from the standpoint of water-tight integrity. This added to our anxieties, as it meant that little time would be afforded to save life in case a torpedo found its mark.

The *Maumee* (Lt. Comdr. H. C. Dinger), an oil tanker, sailed from Boston a few days before the expedition sailed from New York to take her previously assigned station on the route of the convoy to refuel the destroyers as might be required.

Oiling at sea was one of the maneuvers which had been developed in the Destroyer Force three or four months before the war. A division of destroyers had been oiled en route to Queenstown at the rate of 35,000 gallons per hour, in a moderate sea, and the wind blowing a half gale. Without the ability to oil at sea the destroyers would have had to be towed and the eastward movement correspondingly delayed. Only the newest destroyers, those which could get over to the other side by one refueling, were designated to go all the way across, while the old boats, the short-legged fellows, as they were called, went only half way or as far as their oil could carry them, and then returned to New York, or in case of necessity called at St. Johns or Halifax, and as a rule they had to steam against strong headwinds on the way back.

The work of converting the requisitioned cargo ships was pressed to the utmost. They were armed with guns, fitted with lookout stations, a communication system and troop berthing accommodations. The method of commissary supply and messing was worked out and the sanitation of the ships improved as far as possible. Life belts were supplied in a quantity 10 per cent in excess of the number of passengers carried. Special measures were taken to protect life in case of casualty, and sufficient rafts were provided so that if life boats on one side could not be launched because of the listing of the ship or other reason, all hands could still be accommodated. Attention was given to the paramount necessity of landing the troops in good health and in good spirits. The instructions issued to all ships were, in brief, as follows, and every man had to be as familiar with them as with the Lord's Prayer:

1. The use of maximum speed through the danger zone.

2. Trained lookout watches made effective by an efficient system of communication between officers of the deck and fire control watch.

3. Continuous alert gun watches in quick communication with lookouts through the fire control officer.

4. Constant zigzagging.

5. Minimum use of radio; reduction of smoke to a minimum; darkening of ships at night; throwing nothing overboard lest it point to the trail.

6. A trained officer always alert and ready to use the helm to avoid torpedoes.

7. Special prearranged day and night signals between ships on manner of maneuvering when submarines were sighted.

8. Use of guns and depth bombs by all transport and escort vessels.

In addition, it was directed that Abandon Ship drills be held daily; that in the danger zone at daybreak and twilight, the hours most favorable to submarine attack, troops be assembled at Abandon Ship Stations fully equipped and prepared to leave the ship; that watertight doors always be kept closed; that all communication pipes and ventilator ducts be kept closed as much as possible; that the water-tight bulkheads be frequently examined — in short, that everything possible be done first, to guard against disaster, and second, to save the ship and to save life if mined or torpedoed.

On the 4th of June, I went to Washington for final instructions. On leaving the Secretary's office, Mr. Daniels said: "Admiral, you are going on the most important, the most difficult, and the most hazardous duty assigned to the Navy — good-by." My friends in the Department wished me God-speed and that night I returned to New York.

On June 7, 1917, I issued the following secret order to the Commanders of the ships comprising this First Transport Fleet:

U. S. S. *Seattle*, Flagship

Op. Order.
No. 1.
7 June, 1917.

FORCES:

(a) Convoy Group One.
  Train: *Tenadores, Saratoga, Havana, Pastores.*
  Escort: *Seattle, Corsair, De Kalb, Wilkes, Terry, Roe.*

(b) Convoy Group Two.
  Train: *Momus, Antilles, Lenape.*
  Escort: *Birmingham, Aphrodite, Fanning* (if ready), *Burrows, Lamson.*

(c) Convoy Group Three.
  Train: *Mallory, Finland, San Jacinto.*
  Escort: *Charleston, Cyclops, Allen, McCall, Preston.*

(d) Convoy Group Four.
  Train: *Montanan, Dakotan, Occidente, Luckenbach.*
  Escort: *St. Louis, Hancock, Shaw, Ammen, Flusser, Parker.*

1. Reports of enemy submarine activity indicate that the area of greatest activity is East of longitude twenty West, and within a circle radius five hundred miles from FAYAL, AZORES. Submarines may be encountered on the Atlantic Coast of the United States and Canada. Every effort has been made to hold secret the sailing of the convoy but it may be assumed that the departure of convoy from the United States and the hour of departure will be communicated to the enemy. It is possible that particular effort will be made by the enemy to accomplish the destruction of the convoy and no part of the waters traversed by the convoy may be assumed to be free from, submarines. Enemy raiders may be encountered.

2. This force will furnish transportation and escort for U. S. Troops and their equipment to the port designated.

3. (a) (b) (c) (d). Escort assembles AMBROSE CHANNEL LIGHTSHIP at ..... — see area clear. Convoy arrive AMBROSE CHANNEL LIGHTSHIP at Groups take formation specified and proceed without delay on course and speed signaled by Group Commander.

The date and hour for departure will be given orally to Group Commanders.

(v) Coal burning destroyers attached to any group will continue with Group such distance as will permit destroyers to return to U. S. Port without refueling— where they will report arrival to Commander in Chief and request instructions.

(w) The *Maumee* has been stationed at sea on the route of the convoy to refuel oil-burning destroyers. When last destroyer of Group Four has refueled *Maumee* will proceed to ST. JOHNS, NEWFOUNDLAND. Group Commanders will be informed orally of Maumee's position and this position will also be contained in the sealed instructions to each ship.

(x) Oil-burning destroyers will refuel from *Maumee* when directed by the Croup Commander. Seven hundred fifty ton destroyers will after refueling from *Maumee* continue with convoy such distance as will permit them to proceed to ST. JOHNS, NEWFOUNDLAND, without again refueling. Thousand ton destroyers will continue to destination.

(y) Yachts will continue with convoy to destination. If it become apparent that their fuel capacity is not sufficient, yachts will be directed to proceed to FAYAL, AZORES, to refuel and thence to destination.

(z) Before arrival at destination convoy will be met by a Division of destroyers. This division of destroyers will form part of escort from meeting point to destination.

4. All ships will be fueled to maximum capacity.

5. Convoy Commander in *Seattle.* If necessity for use of radio arises use sig code quintuple cipher 3084 — 672—5934—186—7865.

(s) D. W. Blamer, Captain, U. S. N., *Chief of Staff* for Albert Cleaves, Rear Admiral, U. S. N. *Commander Destroyer Force, Atlantic Fleet and U. S. Convoy Operations in the Atlantic.*

Copies to: Operations, C-in-C. Ships Mentioned (2).

Previous to sailing, two groups were assembled in the North River and the others at the overcrowded anchorage at Tompkinsville, Staten Island. At daylight on the 14th of June, the ships got under way in one of the densest fogs that I have ever seen in New York. This called for good seamanship, but the movement in the narrow tide-swept channel was accomplished without mishap, save for one destroyer, which was caught in the anti-submarine nets at the Narrows and had to return to the Navy Yard for repairs.

When this was reported to me, I recalled the fact that just before sailing the Captain of this destroyer handed me a report to the Commandant of the District stating that in his opinion the submarine net at the entrance of New York Harbor was inadequate and not likely to stop any determined enemy submarine which might attempt to enter. At about the same time that the Commandant received this report the author appeared in the Commandant's office with his request for repairs to damages caused by that very net.

This was not the only trouble which developed. The *Corsair* was unable to keep up owing to green firemen and fell back to the second group, being replaced by the *Fanning,* which afterwards greatly distinguished herself by capturing the first prize of the war. Another destroyer developed condenser trouble on June 16th and was sent back to New York for repairs.

The groups sailed at intervals of two hours from Ambrose Channel Light-ship, except Group IV, which was held by the Department twenty-four hours for belated dispatches and stores. Group I was the fastest, Group IV the slowest, and their departure was timed to avoid congestion at the eastern terminus. It is obvious that as the expedition advanced the intervals between the groups opened out, thus increasing the difficulties of submarines lying in wait to attack.

The first group proceeded at fifteen knots, the second at fourteen, the third at thirteen and the last at eleven.

Throughout the voyage the weather was pleasant. The morale of the troops was excellent, and as they familiarized themselves with the ship's organization and plans for saving life in case of disaster, their confidence in those responsible for their safety overcame the natural timidity of the landsman embarked on such a dangerous passage.

The necessity of having no one on lookout who could not speak the language soon became apparent. In the inky blackness one night I asked a muffled lookout on the quarter-deck what he would do if he saw a periscope. He replied: "I gotta tell 'a Sargn't."

He was one of the bandsmen.

At 10:15 P. M. June 22nd, in Latitude 48° 00' N., Longitude 25° 50' West, the first group was attacked by enemy submarines.

At this time we were crossing the line from North Ireland to the Azores, the probable route of U-boats bound for those islands. The Azores were then under suspicion as hiding an enemy submarine base. The escort reenforcement from the east had not joined up but was due to meet us a little further to the eastward. It is suspected that the Germans also had this information and timed their attack accordingly. After the sinkings made by the *U-53* off Nantucket, however, we knew that we had to be ready at all times, even in our home waters, to repel submarines, and we were not caught off our guard. The cruising dispositions made for protecting the troop ships placed the cruiser *Seattle,* my flagship, ahead, and to starboard of them, with the auxiliary cruiser *De Kalb* ahead and to port, thus covering the most likely angles for enemy approach.

My first intimation of the presence of the enemy was the report by *Seattle* lookouts of sighting in the extremely phosphorescent water the wake of a submarine crossing our bow from starboard to port toward the convoy. Simultaneously with this report and before the *Seattle* had time to give the prescribed alarm, the *De Kalb,* which had sighted two torpedo wakes, one passing ahead and the other astern, opened fire. Two torpedoes also passed close to the *Havana.* Captain Gherardi, of the *De Kalb,* who was on the bridge of his ship, told me afterwards how he plainly followed for several hundred yards the unmistakable straight track of a torpedo by its telltale wake of bubbles. He handled his ship to perfection and disaster was averted. It is the irony of fate that little is heard of the Captain who by constant vigil and alert attention to duty saves his ship. It takes an explosion with attending loss of life and excitement to make popular appeal. This, however, is beside the point.

Fortunately, our protective measures were effective and in accordance with the prescribed doctrine the maneuver to evade the enemy at night was performed in a prompt and seamanlike manner. The ships of the right and left columns of the convoy turned to starboard and port, respectively, and ran at full speed as per instructions. There were no torpedo hits and no evidence of injury to the enemy. The convoy reformed at daylight and proceeded on its course.

Lieutenant T. E. Van Metre, U. S. Navy, Executive Officer of the destroyer *Wilkes,* was stationed at the special listening device known as the oscillator which had recently been installed in that vessel and he reported: "I put on the headgear and heard coming into the receivers quite strongly the sound of a submarine running totally submerged. The sound grew in intensity until I could hear it in both receivers, although much louder in the starboard one. I called to the Captain (Lieutenant Commander J. C. Fremont, U. S. Navy) that I believed the submarine was close aboard on the starboard bow."

A little later, after we had left this submarine astern, Van Metre reported that he heard faintly what he believed to be another submarine running on the surface. This experience demonstrated in a striking way the value and possibilities of listening devices to detect submarines.

It was afterwards learned from a confidential bulletin issued by the French Admiralty on July 6th, that on the 25th of June in the same vicinity the British steamer *Fernleaf* was attacked, and on the 29th of June, 400 miles north of the Azores, the *Benquela* and *Lyna* were sunk. On the 4th of July, presumably from the same submarine, the port of Ponta Delgada in the Azores, received a few shots. It was the conclusion of the French bulletin from these activities that submarines had been sent out from the German secret base known to exist in the Azores, to intercept our convoy expedition.

On the 24th of June, in the afternoon, we sighted the Queenstown destroyers, and right glad we were, too, as they bore down to take their stations around us. We had joined up with them on time at the appointed rendezvous, which was a good piece of navigation on both sides. The destroyers immediately began zigzagging and I remember that Hanrahan's swift destroyer, the *Cushing,* took station ahead of us and seemed to be doing a sort of "do-see-do" figure of the old-fashioned quadrille.

The next day smoke was sighted to the northward and I sent Alford Johnson, commanding the destroyer *O'Brien,* to reconnoiter. It proved to be the French escort, composed of two small destroyers.

Floating wreckage all around us gave sinister evidence of the activities of the submarines as was also frequent S. O. S. and "Allo" wireless calls. At this time the submarine warfare was at its height and it has been frequently remarked that never after was there so much wreckage to be seen in the Bay of

Biscay. At 3:00 P. M. we sighted Belle Isle and shortly afterwards two airplanes and a dirigible were seen over the land coming out to greet us.

The most trying experience of the voyage was now to come. We were passing to the southward of Belle Isle when at 8:00 P. M. I received a radio report from Brest, dated at 6:00 P. M. that two submarines had been sighted thirty miles to the southward of the group's then position, both submarines standing to the northward. In other words, they could easily have been within six miles of the group at the time the message was received.

In spite of this all vessels of the group were compelled to slow down well outside the entrance to Croisic Bay, in order to take on board pilots to steer us through the mine fields which the Germans had generously planted in the vicinity, and I confess to a bad quarter of an hour while waiting to get them on board. So much time was lost that the group was compelled to anchor in the open roadstead and wait for daylight.

That the group was not attacked that night seems inexplicable as the Germans lost a marvelous opportunity. As a matter of fact, the channel we took into St. Nazaire was planted with mines by the enemy the following night, and the other groups of the expedition used another channel.

The fourth group was also attacked on June 29th, and the Commanding Officer of the *Luckenbach* reported:

"About 10:30 A. M., this vessel was attacked by a submarine, and one torpedo was seen to pass within about 50 yards of the *Luckenbach*. The course of the ship had just been changed by the Commanding Officer to avoid this torpedo, and the torpedo was seen to come to the surface in the wake of the *Luckenbach* at the point where the change of course took place."

Of this incident the Commanding Officer of the *Kanawha* reported that he saw a submarine when the torpedo was fired and watched the wake, the submarine being directly between the *Kanawha* and *Luckenbach,* and close upon the Kanawha. It was his opinion that the *Luckenbach* would be hit and his crew gave a cheer when they saw her change course to go clear.

Lieutenant (j. g.) J. C. Carey, U. S. Navy, was in charge of the battery of the Kanawha, which did some very good shooting at this submarine's periscope, dropping some shells almost on top of it, if they did not actually hit this small target. It may well be that those shots so confused the aim of the submarine as to cause her torpedoes to miss.

Commander A. C. Pickens of the *Luckenbach* later remarked that he was just as well satisfied that the torpedo missed because his cargo consisted of 5,000 tons of ammunition, with a deck load of gasoline, hay, motor oil, and oxy-acetylene tanks.

The second group encountered two submarines, the first at 11:50 A. M. 26th of June in Latitude 47° 01' N. and Longitude 6° 28' W., about 100 miles off the French coast, and the second two hours later. The group was under escort of six additional American destroyers at the time.

Both submarines were successfully evaded, and the destroyer *Cummings,* upon sighting the second submarine, headed for it at twenty-five knots. The submarine immediately submerged and the periscope was lost to view, but the course of the submarine was plainly disclosed by a wake of bubbles. The *Cummings* passed about twenty-five yards ahead of this wake and dropped a depth bomb, the explosion of which was followed by the appearance of several pieces of lumber, oil, bubbles and debris upon the surface. There was no further evidence of the submarine, and if not destroyed it is probable that it was at least badly damaged.

Commander Neil, who made the counter attack on the submarine in the *Cummings,* was decorated by the British Government for this exploit.

On the 2nd of July the last group arrived at St. Nazaire. The entire expedition crossed without the loss of a man; one officer reported: "We didn't lose but one horse, and that was a mule."

The German Admiralty had boasted that not one American soldier should set foot in France. The bluff had been called and it could not have been called at a more psychological moment.

Commander W. R. Sayles, our Naval Attaché in Paris, was charged with the important duty of arranging the landing at St. Nazaire. Secrecy was of first importance and conditions in France at that time made this particularly difficult.

Commander Sayles made his plans as though the convoy were going to Brest. As he expected, the Germans found this out and in the belief that our ships were bound for that port, the enemy mined the entrances. The sinking of the French cruiser *Kleber* by one of these mines is grim evidence of what had been prepared for the Yankee troopships.

It is not surprising that the Germans were deceived as to our destination. Brest is an admirable port for troop debarkation, whereas St. Nazaire is ill adapted for this purpose by reason of its small size and lack of facilities. This port was not well known and the landing of the first expedition put it on the map.

The channel is of such depth that vessels of deep draft can enter only at high water, while a five-knot tide makes it hazardous for vessels to move inside the harbor except at slack water. The holding ground is also bad. At various times our ships dragged anchor and serious damage was narrowly averted.

The fact that little or no preparation was made there for our reception no doubt helped to deceive the enemy but also had disadvantages as will be seen.

The arrival of so many transports within so short a space of time caused great congestion in St. Nazaire's small harbor. All the troopships carried cargo and large quantities of troop equipment and stores. The unloading of all vessels and quick preparations for the return voyage presented a perplexing problem with the poor facilities available and the shortage of labor.

Five hundred negro stevedores had been brought from the United States by the Army to discharge ships, but they were found altogether unequal to handle such a large number of vessels. The Marine Regiment, which had been carried in the *Henderson, De Kalb* and *Hancock,* relieved the situation somewhat by turning to and discharging their own vessels.

The sea wall was a scene of great activity as the docks, cranes, and railroads, endeavored to absorb several hundred per cent more than their usual capacity. From the ships' holds were discharged boxes of provisions, ammunition, locomotives, baled hay, horses, automobile trucks, gasoline and other Army impedimenta. French stevedores, American sailors and marines, negroes and German prisoners worked side by side.

The basin at St. Nazaire was something to look at for the next two weeks, but one to which the inhabitants of that quaint Breton town were soon to become accustomed.

From the transports an almost continuous stream of troops marched off over the cobblestones of the narrow winding streets to the camp in course of construction by German prisoners a few miles behind the town.

The population gathered along the quays looked on in whispering wonderment at the young khaki-clad strangers who had appeared, almost overnight, from over the seas. There was no cheering, no patriotic demonstration, only the respectful silence of the women and children, the old men and the broken soldiers. In their eyes, however, was unuttered thanksgiving and also an unconscious, wistful look to see what they could read of the spirit of America in the faces of these, her soldiers.

It was a joining of hands in war of France and America recalling the days of Lafayette and Rochambeau. In a few days the shyness disappeared and in its place came occasional smiles and spoken greetings. The message from America had been read aright — it was the promise of a great nation to stand by France to the finish.

Chapter Three - The Stay in France — The Return Voyage

On the evening of our arrival at St. Nazaire, I dined at La Boule, an attractive resort on the coast, a few miles out of town, and learned that General Pershing was to make a visit of inspection the following day to the new United States Army camp in process of construction just outside the city.

It was arranged for General Pershing and General Sibert to lunch with me on board the flagship *Seattle,* and I invited a number of American and French officers to meet him. Major Frederick Palmer, at that time attached to Pershing's Staff, suggested that a number of French newspaper correspondents be asked to come on board after the luncheon for the purpose of interviews and taking photographs.

The cabins of the *Seattle* were filled to overflowing and every one was in the best of humor. I remember what a fine impression the American Commander-in-Chief made upon every one. He was accompanied by his personal aides and by General Peltier of the French Army, who had lost his arm at the Marne and had been detailed to Pershing's Staff by the French Government.

After luncheon the newspaper men, about thirty, came into the cabin where I received them and presented them to Pershing in a brief speech in which I spoke of the services of the French Fleet in the Chesapeake which made possible Washington's victory at Yorktown in 1781. In acknowledging this essential aid rendered by France to us in our struggle for Independence, I said it was particularly gratifying to have the honor of commanding our first troop convoy expedition to French shores. The General also made a few remarks and we both went on deck, where many pictures were taken in a pouring rain.

It was at this time that Joseph Dunn, the stern man of the Captain's gig, was washed overboard, and before the boats could get to him he was swept away and under by the swift current. This was the first death in the Force. A few days later his body was recovered and shipped to the United States in the U. S. S. *Cyclops,* the ship which later in the war so mysteriously disappeared. [1]

We had another man-overboard-episode while in St. Nazaire which ended more happily. One evening, while the returning sailors of one of our liberty parties were embarking in the *Seattle's* boats at the dock, one of the party reported seeing a man overboard and going down. He jumped in and dove several times but could not find anything but a neckerchief. When the party was mustered in, our sail-maker's mate, old Ben Amble, was the only one reported missing and as the name on the neckerchief appeared to be his, he was supposed to have been drowned.

[1] U. S. S. CYCLOPS, *Mystery Ship.*

The 19,000-ton naval collier *Cyclops* disappeared at sea mysteriously, having been last heard from on the 4th day of March, 1918. She was a large and most valuable vessel carrying a crew of 293 officers and men. Quite some time afterward a bottle floated ashore at Baltimore containing a note signed by "John Rammond" of Chicago which said:

"Our ship, the *Cyclops,* has been captured by an enemy submarine."

There was, however, no man of the crew by the name of Rammond and the Navy Department attached no truth to this note found in such an unlikely locality.

Enemy submarines had been frequently reported operating in the West Indies waters, and one theory advanced was that during the *Cyclops'* last stay In port before March 4th some German agent delivered to her forged orders which may have led to her capture and destruction. The ship carried a valuable cargo of manganese. All theories of the destruction of this vessel were carefully traced by the Navy Department, but no evidence has come to life to indicate her fate and the ship has Joined the ranks of the mystery ships of the sea.

The next morning a diving party went over to the dock to search for the body and while a diver was down and other men dragging the bottom with grapnels, who should come calmly ambling down to the dock but old Ben Amble himself. When he learned what all the fuss was about and that he had been reported drowned, he was a surprised man and remarked that that was the first he knew of it.

When the work of debarkation from the transports was finished, at the suggestion of our Embassy, I went to Paris for the Fourth of July celebration. I arrived there on the morning of the Fourth, accompanied by my two aides, Lieutenant Commander A. L. Bristol and Lieutenant T. A. Symington. We were met at the Quai d'Orsay by our Naval Attaché, Commander Sayles, and by a French Naval officer representing the Admiralty. The latter informed me that he had been appointed my personal aide and liaison officer during my visit to Paris, and also placed a car at my disposal with the compliments of the Ministry of Marine.

The Fourth of July, 1917, in Paris, was an eventful day filled with historic incidents. The first function of importance was the presentation of a stand of colors to the United States Army by a French society. This took place at the "Invalides."

General Pershing received the colors in the presence of President Poincaré. The court was crowded and every one seemed thrilled by the presence of the American troops and, indeed, it was a stirring ceremony and one which must have made a lasting impression on all who took part in it. There were many war trophies from the recent battlefields in the enclosure and over all stood the heroic statue of Napoleon, looking down, as it were, from the top of the Hotel des Invalides.

After the ceremony my own party drove through cheering crowds to the Admiralty Building, where I was presented to the Minister of Marine, Rear Admiral Lacaze, and to the Chief of the Naval General Staff, Vice-Admiral Le Bon. This visit was necessarily brief because we had to hasten from there to the Cemetery of Picpus to attend the annual ceremony of the American Society in Paris at the grave of Lafayette.

Addresses were made there by Colonel Stanton, General Pershing and Ambassador Sharp, and it is my distinct recollection that it was Colonel Stanton, who, in his eloquent address, exclaimed: "Lafayette, we are here!" Many of the most distinguished officials in Paris were present, among them Marshal Joffre and Mr. Brand Whitlock.

I was much impressed by the tremendous enthusiasm shown by the bystanders who crowded the streets as our troops marched by. Young girls and women rushed into the streets and showered flowers upon them. Many flowers were thrown into my car.

Luncheon at the American Chamber of Commerce in the Quai d'Orsay was attended by several hundred people. Speeches burning with enthusiasm and patriotism were made by M. Ribot, M. Viviani and others. General Pershing spoke briefly, but to the point. In M. Ribot's address he said, with visible emotion, "This is not only the Independence day of the United States, but it is the Independence day of the Nations."

This luncheon was followed by an official reception given by Ambassador and Mrs. Sharp. The Embassy was thronged and all were in good spirits. It was quite evident that the pendulum had swung the other way and that the safe arrival of the American ships with troops had inspired the city with a gayety that it had not known for many months.

In the evening General Foch gave a dinner to General Pershing and his Staff at the Armonville in the Bois. To meet him were invited the Minister of War, the Military Governor of Paris and five Generals of Divisions.

I sat on the right of M. Painleve, the Minister of War, and he spoke in the most commendatory terms of the safe transportation of the first American troops to France. Mr. Whitlock had said practically the same thing to me at Picpus, and also added that the American Navy had written a new page in history.

I shall never forget the welcome that those warm-hearted people gave to us all. Words cannot describe it. It showed the tremendous moral effect, even upon the man in the street, which the safe arrival of the first convoy at this critical period had made upon the nation.

THE RETURN VOYAGE

It was necessary to return all the ships of the expedition to the United States at the earliest possible moment. As soon as their cargoes were discharged, the troopships were dispatched home in groups, escorted by the cruisers and destroyers which had come over with us, reinforced across the zone of greatest submarine activity by destroyers from Queenstown. The latter accompanied them to about 600 miles off the coast and then returned to Ireland.

I remained on board the *Seattle* at St. Nazaire until the last ships were cleared and at noon of the 14th day of July sailed for New York, escorting the *Cyclops, Kanawha, Occidente, Luckenbach, Dakotan, Momus* and *Montanan.*

At my request, the French Government had made the preparations for our return escort through the submarine zone much more elaborate than had been the preparations for our arrival. The reason for this was that during our stay in St. Nazaire the submarine activities had greatly increased. No less than three large American schooners had been torpedoed near the mouth of the Loire and two steamers had been mined near Belle Isle. Our destroyer escort consisted of three French destroyers and five of our own. The French Government also supplied a dirigible and one or two airplanes. Two mine-sweepers preceded us.

The rapid expansion of the Navy meant that we had many new recruits to train and careful attention was given to the instruction and drill of gun crews. In order not to delay convoy operations a method of conducting target practice en route was devised during the first expedition and all ships had target practice at sea when clear of the zone of greatest submarine activity.

A periscope target was designed which would dive and expose itself at irregular intervals when towed 300 yards astern. All vessels were equipped with these targets, and each ship also organized a so-called rake party to take station in the stern of the towing vessel and observe the splashes over a long graduated rake measuring and recording the distance of the splashes over or short of the periscope target.

The accuracy of the firing ships was thus checked and a method of scoring having been determined the gun crews were in competition, and excellence encouraged by the award of prize money.

The firing ship was required to maneuver as it would in actual torpedo attack, heading for or away from the periscope. To hold practice it was not necessary to stop the convoy. The towing ship and firing ship were designated by signal and proceeded with the firing, the other ships continuing on their course and keeping clear.

Keen interest was taken in this target practice and it is probable that more than one transport was saved by the skill developed in this manner by the gun crews in dropping a shower of shells near, if not on, the periscope of an attacking submarine, thus confusing the enemy's aim.

Later the ships were supplied with diving or plunging shell designed to follow an underwater trajectory and to explode on contact with, or in the vicinity of, the submerged submarine. The principle of the diving shell was the same as that of the depth-bomb and they no doubt added to the embarrassments of the U-boat commanders. The details of these and other anti-submarine devices will be described in a later chapter.

The return voyage was uneventful with the exception of a ripple of excitement one afternoon when we thought we had sighted the German raider *See Adler.*

Before sailing from St. Nazaire, I had obtained from the French Admiralty the latest information of this vessel, and she was described as follows:

"The German raider *See Adler,* now probably operating in the Atlantic, was formerly the full rigged American ship *Pass-of-Balmaha,* about 1,500 tons, steel hull, built at Glasgow in 1888. (See sketch attached.)

"The *See Adler* is an auxiliary, both sail and steam, and is driven by a Diesel engine, giving a speed of about 11 knots in fine weather. In a strong favorable breeze she can make 16 knots.

"Her armament probably consists of four 6-inch guns, two concealed on each side; four 4-inch guns, one on each bow; two movable machine guns, and a range finder mounted on the forecastle. She also probably carries four torpedo tubes and 75 mines.

"When a ship is in sight canvas cowls are rigged; a man in feminine dress, carrying a sunshade, is often seen on the poop."

The strange vessel sighted by the *Seattle* was a three-masted square rigger, closely resembling this description. She acted so suspiciously that I directed that a one-pounder be fired across her bow to bring her to. This had no effect and was followed by a 3-inch shell with better result. The *Seattle* then approached the suspect with caution, maneuvering to keep out of possible torpedo range. All our guns were trained upon her ready to open fire instantly and one 6-inch shot from our secondary battery would have blown her out of the water.

Much to our disappointment, as we closed up, she displayed the English red ensign and proved to be a Newfoundland fisherman. It must have been from some such craft that the Germans copied the *See Adler's* rig as the two were almost identical.

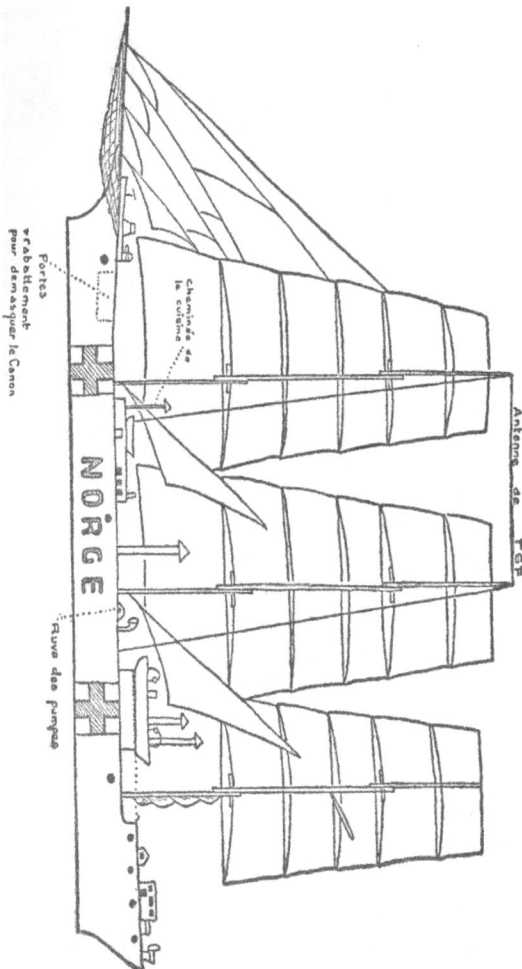

All ships made the return voyage in safety and I had the infinite satisfaction of receiving letters of congratulation from General Sibert, who commanded the troops of the first expedition, from the Secretary of the Navy, and from the Secretary of War. All of these were published to the entire force and the commendation was a great spur to further endeavors.

LETTER FROM GENERAL SIBERT

"Headquarters, First Expeditionary Division, France, July 2, 1917.
Commanding Officer, U. S. Naval Convoy,
First Expeditionary Force.
My Dear Admiral Gleaves:

The safe arrival this date of the fourth and last division of the first convoy, prompts me to convey to you my sincere congratulations upon the successful completion of the difficult task with which you were charged.

In as far as I can speak from personal observation and from hearsay, I desire also to express my appreciation of the highly courteous treatment which the Army invariably received at the hands of your subordinates in the Navy charged with duty on board of the transports.

I am Very sincerely yours,

(Signed), Wm. L. Sibert, Major General, U. S. Army.

LETTER FROM THE SECRETARY OF THE NAVY
NAVY DEPARTMENT, WASHINGTON, D. C.

4 August, 1917.

My Dear Admiral:

I have received the official report containing the details of the several attacks made by submarines upon the ships under your command, carrying the first American troops to France, to take part in the war. I have read this report with the deepest interest and have sent exact copies of it to the House and Senate Naval Affairs Committees. I have also given out to the press a copy, omitting the names of the ships and latitude and longitude in which the attacks took place. I wish to express the appreciation of the Department, and also of the whole country to you as Commander, and to the officers and men to whom was committed this hazardous and important undertaking. It is a matter of national rejoicing that the troops arrived safely and that you executed this important duty in a manner to call for the highest commendation.

Sincerely yours,
Josephus Daniels.

On July 3, 1917, the Secretary of War wrote the following letter to the Secretary of the Navy:

My Dear Mr. Secretary:

Word has just come to the War Department that the last ships convoying General Pershing's Expeditionary Force to France arrived safely to-day. As you know, the Navy assumed the responsibility for the safety of these ships on the sea and through the danger zone. The ships themselves and the convoys were in the hands of the Navy, and now that they have arrived and carried, without the loss of a man, our soldiers, who are the first to represent America in the battle for democracy, I beg leave to tender to you, to the Admiral, and to the Navy, the hearty thanks of the War Department and of the Army. This splendid achievement is an auspicious beginning and it has been characterized throughout by the most cordial and effective cooperation between the two military services.

Cordially yours,

(Signed), Newton D. Baker,
Secretary of War.

Chapter Four - Lessons Learned from Experience of First Voyage —Repairing the German Ships

Many valuable lessons were learned from the experiences of the first voyage and steps were immediately taken to incorporate them in the development and expansion of troop transportation work.

In my mind a most important lesson taught by this voyage was that the transportation should be done entirely by the Navy, and I believe further that this was the unanimous opinion of all the army officers with whom I discussed the subject.

A method of procedure was agreed upon by the War and Navy Departments and having been approved by the President had all the force of statute law. Charter rules governing the Army and Na\y in convoy operations were set forth in a confidential order signed and promulgated by the President as Commander-in-Chief.

The task of protecting military expeditions embarked on the sea is purely naval and many of the most important measures of protection in submarine waters are those which must be enforced within the transport itself.

The Navy was the establishment best equipped for quickly organizing and operating the transport fleet. To this view the success of the first expedition added weight.

Upon my return I was called to Washing-ton in conference with the Secretary of Navy and the Chief of Naval Operations. I strongly urged that the operation of the transports be taken over entirely by the Navy and that they be fully manned by Naval officers and crews. Shortly after, the War and Navy Departments jointly recommended this plan.

The dividing line of authority in the transport service was made at the docks; the Army superintended the docks in the ports of embarkation and debarkation, providing and loading passengers and cargo; the Navy took charge afloat, provided and routed escorts and convoys, manning, operating, repairing, coaling and provisioning the transports.

Providing a transport fleet was pioneer work. Ships had to be obtained, officers and crews enrolled and trained. It was necessary to have docks, storehouses, lighters, and tugs, coaling equipment, repair facilities, and all the varied machinery for operating and maintaining a large transportation service. An efficient administrative organization had to be developed and red tape had to be cut.

During the first voyage we also learned a great deal which proved useful in developing a sound doctrine. It is always the unexpected which happens at sea, especially when fighting submarines, and it was my policy and endeavor never to restrict any Captain by hard and fast rules. He was always encouraged to use his own discretion and was given the assurance that in doing so he would always have my backing and support.

The best protection of a transport from torpedo attack is alert seamanship. In this our Captains excelled. Theirs was not a spectacular position and few people appreciated the weight of responsibility they carried and the strain of their constant vigilance. Their reward is satisfaction in difficult and important duty well done.

The first experience broadened our ideas and views of the entire subject. I summoned the Captains to frequent conferences and frank discussion cleared the air. Finally a set of orders, confidential, special, and general, were developed which taken together made an organization flexible but thorough and practical in every way.

One rule, however, I emphasized from the beginning and it was hard and fast — in case a transport of a convoy was torpedoed the other troopships steamed away at full speed and left the rescue work to the light draft escort craft. Early in the war it cost England hundreds of lives and three fine cruisers, the *Aboukir, Cressy* and *Hogue,* to learn the lesson that to go to the assistance of a torpedoed ship is to play into the hand of the lurking U-boat. We profited by her experience and lost no ships in this way. Our Captains obeyed the above rule scrupulously, although it went strongly against their instinct, which was always to go to the assistance of a ship in distress.

As soon as the number of transports in service permitted, the policy was adopted of sending them in groups composed as far as possible of not less than four nor more than eight vessels, all of about the same speed, each group escorted by a cruiser and two destroyers and sailing at intervals of eight days. Rendezvous at sea were established with destroyers on the other side to escort the troopships through the danger zone of greatest submarine activity.

The transportation of the Army to Europe was a joint Army and Navy proposition, and it could not have been handled satisfactorily had it not been for the unity with which the services worked together. The War Department was represented in Hoboken, New Jersey, which was the principal home terminal, by Major General D. C. Shanks, U. S. Army, and at Newport News by Major General Grote Hutchinson, both men of large views and broad-gauged ideas. Rear Admiral H. P. Jones, an officer of rare judgment and ability, commanded the Newport News Division of the Cruiser and Transport Force, The cost of the transportation and the expenses of upkeep, repairs and maintenance, were paid by the Army.

Thus it will be seen that the overseas movement was by no means a one-man task. There were many engaged in it, and the success of it is due to the fact that the Administrative and Executive heads worked together on shore, and those of the Force at sea faithfully, efficiently and zealously executed their orders.

The destroyers were enabled to perform transatlantic escort duty by the stationing of a tanker in mid-ocean from which they could refuel.

The first oiling at sea of our destroyers en route for Europe was done on May 28, 1917, when six of the oil burners bound for Ireland were enabled to make the trip under their own power. I was keenly interested in "oiling at sea," as the operation had been developed in pre-war days under my supervision, so I sent my personal aide. Lieutenant Commander Perkins, in the tanker *Maumee,* detailed to this duty. The following is quoted from one of his letters written to me from the *Maumee.*

The first part of the job is over and was successfully accomplished, although the weather was very unfavorable.

There was a heavy sea running and a fresh breeze blowing, but by making a lee we could take one at a time and finished the six in one day.

Due to the heavy seas we parted two or three hawsers but there were no accidents and no damage.

The night after we finished (May 28, 1917) it blew up a gale from the northwest which lasted until this morning (May 31st).

We make port (St. Johns, Newfoundland) tomorrow and will be ready to take out the next lot when they arrive but so far we have not been able to raise them (by radio).

It is quite cold up in this part of the world. We passed a number of icebergs — one big fellow this afternoon.

Destroyers took an essential part in our transportation work and they never failed us. For over a year before the war I had commanded the destroyer force of the Atlantic Fleet and as I retained with me my old Staff when assigned convoy duty, the experience we had gained with the destroyers contributed to the close cooperation and understanding which existed throughout the war.

It was at once seen that a properly developed aviation service would prove of great value in troop transportation work, and our Navy proceeded to establish numerous aviation stations along the seaboard of the East Atlantic. The first American fighting force landed in France was a detachment of Naval aviators, and Lieutenant Whiting, who had conducted much experimental flying from my flagship, the *Seattle,* wrote me from Paris on July 6, 1917, that while the British and French recognized the importance of an air service against submarines they were much handicapped by lack of material and personnel because of the pressing needs on the Western front and in the North Sea.

Whiting said in part, "The French have awakened and are now commencing to establish stations along their entire coast from Dunkirk to Bayonne, and only a lack of material and unwounded pilots (they are using pilots from the front who have been wounded and some land planes) has prevented this being done. They need all the assistance we can give them both in material and men — not only pilots but all the types of men necessary to maintain a seaplane station and, at present, to establish these stations. Men, material, equipment must be provided quickly to put down the submarine menace if troops are to be brought over in safety."

The interest of the transportation service in aviation is obvious and it was a happy coincidence that the peace time development of aviation in the fleet should have been assigned to the armored cruisers *Seattle* and *North Carolina*, the *Seattle* then being my Flagship of the Destroyer Force. I believe that our first use of airplanes at sea for military purposes was made when Lieutenant Whiting, operating from the *Seattle*, made a flight over San Juan del Sur, Cuba, for the purpose of observing the movements of the insurgents, and when he carried a message from me at sea to Admiral Mayo, Commander-in-Chief of the Atlantic Fleet, anchored in Guacanayabo Bay, Cuba.

The first radio equipment installed in seaplanes was in those on board the *Seattle*. I recall that it was while experimenting with radio at Guacanayabo Bay, Cuba, that Lieutenant Chevalier, pilot, accompanied by Lieutenant Lavender as radio operator, in one of the C-type of planes, was forced into a nose dive and wrecked, falling from a height of about 300 feet. Lieutenant Lavender was severely injured, both arms being broken, but Lieutenant Chevalier escaped without serious injury. Before leaving Guacanayabo Bay, Cuba, three of the *Seattle's* five planes had been wrecked.

In this connection it is interesting to note that Lieutenant Commander A. C. Read, who made the first transatlantic flight in command of the NC-4, was one of the aviation officers on board the *Seattle* and had his first experience in flying seaplanes from that vessel.

Subsequent experiments in aviation at sea were continued by the *Huntington*. The *Huntington* demonstrated that seaplanes must be protected from the weather and especially from the blast of guns and also showed the usefulness of a kite balloon on sea-going ships as a lookout.

The system of using the kite balloon in connection With seaplanes, afterward successfully developed abroad, was first tried on board the *Huntington* while escorting troop convoy No. 7. The balloon was used to discover the submarine, the plane being kept ready to launch. In a practice test, dummy depth charges were dropped from the air with great accuracy, falling at an average distance of six and one-half feet from the periscope target. Only the removal of seaplanes from the cruisers prevented further use of this plan.

REPAIRING THE GERMAN SHIPS

Upon the Declaration of War the United States Customs Officers took possession of all German ships in United States ports and the larger vessels were designated to be fitted out as troop transports. Their German names, new names, gross tonnage, and fitting out ports, were;

Former Name	New Name	Gross Tonnage
Grosser Kurfürst	Aeolus	13,102 tons
Kaiser Wilhelm II	Agamemnon	19,361 "
George Washington	George Washington	25,569 "
Frederich der Grosse	Huron	10,771 "
Vaterland	Leviathan	52,820 "
Koenig Wilhelm	Madawaska	9,410 "
Barbarossa	Mercury	10,984 "
Prinzess Irene	Pocahontas	10,893 "
Hamburg	Powhatan	10,531 "
President Grant	President Grant	18,172 "
President Lincoln	President Lincoln	18,172 "

BOSTON

Former Name	New Name	Gross Tonnage
Amerika	America	22,622 tons
Cincinnati	Covington	16,339 "
Kronprinzessin Cecile	Mount Vernon	19,503 "

NORFOLK

Neckar	Antigone	9,835 tons
Rheim	Susquehanna	7,797 "

PHILADELPHIA

Kronprinz Wilhelm	Von Steuben	14,008 tons

The *Prinz Eitel Friedrich* had already been fitted out as an auxiliary cruiser and transport at the Philadelphia Navy Yard and renamed the *De Kalb.*

These vessels were first under the Shipping Board, but the work of repairing them was not progressing satisfactorily and I urgently recommended that they be taken over at once by the Navy. Toward the latter part of July they were turned over to the Navy Department and the work of preparing them for sea was pushed to the utmost. The *Leviathan,* the last ship, sailed for Europe on December 17, 1917.

To a man who really loves his ship, malicious injury to her by her own captain seems almost impossible; but the Teutonic mind is utilitarian rather than sentimental, and so, when we went to war, the captains of these ships, acting, no doubt, under instructions from the men higher up, set to with sledge and chisel to wreck and destroy. (See Chief of German Admiralty memorandum.)

Even so, they were stupid and blundered in the job. We were accustomed to attribute to these men a knowledge and ingenuity almost superhuman, and yet they failed to take into account electric welding, to say nothing of Yankee ingenuity, perseverance and skill.

When these ships were turned over to my command, the repairs had already progressed under the personal supervision of Commander Jessop of the Navy, who accomplished a big work in organizing and directing his gangs. He gained much reputation by devising a method of cleaning the *Leviathan's* bottom by divers, which was most important, for she was very foul after her three years alongside the pier at Hoboken, and, as is well known, there was no drydock in the country large enough to take her.

Before these ships were commissioned several naval officers and a skeleton naval crew were ordered on board each of them, to assist and supervise. Daily reports of progress were made, and each week I held a conference on board the Flagship with my Staff and the officers assigned the different ships for the purpose of interchanging ideas and devising ways and means to expedite the work.

The damage done to auxiliary machinery, piping, and fittings by deterioration from lack of care was, in general, even greater than that done willfully. The boilers, the most sensitive and vital part of a ship, had suffered woefully through neglect, and the ships throughout were dirty beyond description.

The naval crews were gradually filled up to strength, and while machinery repairs were going on, they went ahead with scrubbing, scraping, cleaning, painting, disinfecting, and fumigating, to make the ships habitable and sanitary for the troops.

The chief acts of sabotage had been directed against the main engines. As an example, on board the *George Washington,* the high-pressure cylinders of both main engines, both first intermediate pressure valve chests, and the steam nozzles to both low pressure valve chests were wrecked, — large sections of castings having been broken off, evidently by the use of heavy battering rams. The castings of both main circulating pumps were also battered. The two main engine throttle valves with their operating gear had been removed from the ship together with about thirty boiler manhole plates and parts of various auxiliary machinery.

The biggest job, of course, was the work of repairing the main engines. This was most successfully accomplished by electro-welding large cast steel pieces or patches on the parts of the castings which remained intact. This was completed in a few months, whereas to make new cylinders would have taken over a year.

This electric welding was an engineering feat which the Germans had not calculated on. The enemy had broken out large irregular pieces of the cylinders by means of hydraulic jacks. Where these parts had been left in the engine room they were welded back into place, and in cases where the pieces had been thrown overboard new castings were made.

Electric welding is a slow and difficult process and was carried on day and night, Sundays and holidays, to the full capacity of the available skilled mechanics. After each casting had been welded, the cylinders were machined in place, — special cutting apparatus being rigged for this purpose. Finally each cylinder and valve chest was thoroughly tested under hydrostatic pressure. The repairs to the cylinders were uniformly successful. In actual trial they held up perfectly under hard operating conditions and there was not an instance of the welded portion breaking away.

The auxiliary machinery was also damaged but not to the same extent as the main engines. Some of the dynamo engines and ship's pumps were badly smashed, and the castings of the circulating pumps, which supply the cold salt water for condensing the exhaust steam, seemed to be a favorite object of attack. Several main steam line valves and engine throttle valves had been dropped over the side and numerous machinery parts were missing. Electric wiring in some of the ships was cut and the electric leads interchanged.

The work of repairing the ships was attended by various sorts of difficulties. For one thing, there were no plans of the machinery and but few plans of the ships. The machinery was all of German manufacture and missing and broken parts could not be purchased in the market. Everything requiring renewal had to be specially manufactured, and missing parts had, first, to be designed.

In addition to the long list of machinery repairs, extensive alterations were effected, including the installation of thousands of "standees" or bunks; large increases in the bathing and sanitary plumbing arrangements; the enlargement of the galleys and increase of commissary equipment; the installation and equipment of hospitals; the provision of life rafts, boats and life belts for four or five times the normal number of passengers; the installation of guns and ammunition magazines; and scores of other smaller but important changes necessary to permit the great increase in passenger capacity, and at the same time to keep the ships safe and sanitary.

Most of the ships carried only enough coal for one passage across the ocean. Because of the coal shortage in France and the shortage of colliers, it was decided to increase the coal capacity of all transports to enable them to make the return passage without refueling abroad. To accomplish this some of the cargo holds were converted into bunkers. All naval transports were fitted to carry coal or nearly enough coal for the round trip.

TYPICAL GERMAN DESTRUCTION

In some instances the Germans showed originality, but in the main the destruction was similar in all ships. As an example, the following is a translation of a German memorandum found on board the S. S. *Hamburg,* renamed the *Powhatan,* which describes the wrecking done to that vessel.

1. Starboard and port H.P. cylinder with valve chest; upper exhaust outlet flange broken off. (Cannot be repaired.)
2. Starboard and port 1st M. P. cylinder with valve chest; upper exhaust outlet flange broken off. (Cannot be repaired.)
3. Starboard and port 2nd M. P. valve chest; steam inlet flange broken off. (Cannot be repaired.)
4. Valve chest cover damaged, balance cylinder broken. (Cannot be repaired.)
5. Four relief valves from 2nd M. P., overboard — lost.
6. Starboard 2nd M. P. piston guide rod damaged. (Cannot be repaired.)
7. Port 2nd M. P. stuffing-box gland of piston rod guide, overboard — lost.

8. Starboard and port low pressure valve chests: steam inlet flanges broken off. (Cannot be repaired.)

9. Valve chest cover damaged, balance cylinder broken. (Cannot be repaired.)

10. Two relief valves, overboard — lost.

11. Port low pressure stuffing-box gland of piston rod guide, overboard — lost.

12. Port and starboard main engine stop-valve, with by-pass valves and reversing engine valves, overboard — lost.

13. Low pressure relief valves and two guides of valve stem, overboard — lost.

14. Port and starboard exhaust nozzles (outlets) from high pressure to first M. P., three (3) flanges broken off, two relief valves, overboard — lost.

15. First M. P. starboard: exhaust pipe of exhaust line to 2nd M. P., flange broken off. (Cannot be repaired.)

16. Guide of valve stem, relief valve on cylinder, overboard — lost.

17. First M. P. port: exhaust pipe of exhaust line to 2nd M. P., flange broken off. (Cannot be repaired.)

18. Exhaust line to 2nd M. P. damaged, guide on valve stem relief valve, overboard — lost.

19. Starboard and port low pressure exhaust pipes damaged. (Cannot be repaired.)

Translated by,

J. W. Coates,
Chief Machinist, U. S. N. R. F.,
attached to the U. S. S. *Powhatan.*

While the preparation of the big liners was being rushed there was a smaller fleet of vessels steadily pushing its way across and back, carrying a comparatively small number of troops but a most significant promise.

It was not expected that this service could continue without losses. Of the four vessels which sailed from New York on September 24, 1917, two, the *Antilles* and *Finland,* were torpedoed on the return passage. The *Antilles* was sunk and the *Finland* was badly injured but succeeded in returning to Brest under her own steam. These and other losses will be subject matter for later chapters.

Chapter Five - Safeguarding the Troopships

It is interesting to consider some of the new features in organization, equipment and navigation forced upon ships by submarine warfare. The inventor had opportunity to exercise his talent in a field which had no limit. At home and abroad, suggestions for securing the absolute safety of ships came pouring in from well meaning people, who, for the most part, had zeal without knowledge of sea or ship — wonderful fancies for destroying the "sub" and saving the vessel.

For instance, there was one suggestion that upon the near approach of a U-boat, the vessel attacked should fire at the enemy periscope a shell with line attached from a rocket gun, such as is used by the Coast Guard to throw a life line across a ship in distress, the idea being to entangle the U-boat, in other words to lassoo him.

Another suggestion was to make the ship unsinkable by filling her up with water-tight boxes. This was actually tried, but one torpedo sent the alleged unsinkable ship to the bottom.

I think that all propositions, even those obviously impracticable, were given careful consideration. As was to be expected, however, the real defenses against the submarine were devised by practical seamen.

THE TORPEDO

The general characteristics of the torpedo are now pretty well known. It is a highly scientific mechanism consisting of many intricate parts ingeniously assembled in a metal shell about twelve to twenty feet long, eighteen to twenty-one inches in diameter, weighing about one ton, and valued in this country at about $8,000.

In appearance a torpedo somewhat resembles a small, elongated auto-submarine. It has horizontal and vertical rudders which can be so adjusted, in conjunction with an automatic steering device, as to make the torpedo keep at a certain depth and either travel straight or in a curve. The torpedo is propelled by screws driven by an automatic compressed air engine, capable of giving a speed as high as thirty-six knots.

By the act of launching from the tube, a starting lever is tripped, which causes the propelling mechanism to go ahead at full speed. The head of the torpedo carries a powerful bursting charge.

The object of the U-boat is to launch a torpedo so that it will detonate this high explosive against the underwater body of the target ship.

TYPICAL U-BOAT ATTACK

There are any number of variations in the plans used under various circumstances by the different submarine skippers, but for the sake of illustration, suppose a U-boat submerged in a favorable position ahead and slightly on the bow of her quarry, distant 4,000 yards, and approaching to attack at a speed of 6 knots, while the target ship is advancing at a speed of about 12 knots.

As the ship can probably escape by maneuvering if the periscope is seen before the torpedo is fired, the critical time is during the approach.

The problem of the "sub" captain watching through his periscope is not a "cinch." He has to estimate the course and speed of the big fellow, — not an easy thing to do, especially when camouflaged and zigzagging. The U-boat must then be maneuvered so as to be able to "let go" her torpedo just as the target ship passes abeam and close aboard.

41

Beginning at 4,000 yards the submarine can be expected to show about one foot of periscope and observe for a period of about thirty seconds and then disappear. After this, four or five successive observations will probably be taken at intervals of about one minute, the period of time that the periscope is exposed diminishing gradually to ten or five seconds.

When closed to about 1,000 yards or less the firing exposure will be made, and this will probably be for about twenty-five seconds in order to assure a well-aimed torpedo, launched at about 500 yards from its mark.

The above procedure is not absolute — some submarine commanders show more periscope in the approach and others less — but it may be taken as typical. This means that in an average attack, from the time the submarine can be seen to the time the torpedo is fired, about ten minutes elapse, during which there are about fifteen exposures of the periscope for gradually diminishing periods of time, ranging from thirty seconds down to five seconds, except the last exposure for firing, which lasts about twenty-five seconds.

In safeguarding the troopships, the escorting mosquito craft of air and sea - I also have in mind those who laid the North Sea mine barrages — all did wonderful work; but we did not have nearly enough of either destroyers or airplanes to answer adequately the wartime demand. Consequently, the ships of the force I commanded were frequently thrown on their own resources, and as all hands knew that they were on the receiving end of the enemy torpedoes, considerable interest was taken in developing ship defense to the utmost.

The submarine defense within the ship included lookouts, prompt maneuvering with helm, use of maximum speed, guns, depth charges, smoke screens, and camouflage painting.

THE LOOKOUT

The first defense against the U-boat was the lookout. Never has there been so clearly proven the everlasting truth that "Eternal Vigilance is the price of Safety." The old sea phrase of the essentials of safety, the three "l's" - log, lead, and lookout - were all concentrated in one great big "L" during the war.

From the beginning, it was obvious that the entire horizon would have to be kept covered by keen eyes at all times, day and night, and that the usual fashion of instructing lookouts to keep a bright watch on the port bow, on the starboard quarter, etc., would not suffice.

There was no special system established, until, on our first expedition. Lieutenant Commander Gill, the gunnery officer of my flagship, came into possession, in France, of an essay by a captain of the French Navy. From this he developed a practical and scientific method by which every degree of the horizon was under constant examination by keen and tested eyes, watching through binoculars. This system was used by the *Seattle* on her first return voyage and later adopted by the other ships of the force.

No man was kept on watch longer than one-half hour. He was taught never to take his glasses from the assigned arc, indicated by a dial in front of him. Even if a torpedo appeared in another sector, he was still to keep his attention riveted on his own arc, because U-boats sometimes hunt in pairs. Lookouts were intensively drilled until it became second nature to make prompt and correct reports of everything sighted.

In the so-called circle of lookouts, each man had fifteen degrees to cover, so that around one deck there were twenty-four men constantly searching their assigned sectors. Besides these, additional lookouts were stationed alow and aloft.

So, it will be seen why the Navy Department issued its call for voluntary loans of glasses from private individuals; these numerous lookouts required many more than the Navy Department could supply. After the Armistice I was talking to a gentleman, who was one of the volunteers, and asked him if he had ever been thanked by the ship which had received his glasses. He replied, "No, I don't care a damn for the thanks, but I would like to have my glasses back."

Time is everything in a torpedo attack, and the gain of a few seconds in sighting, reporting and putting the helm over may mean saving hundreds of lives.

A striking instance as to how a single lookout saved a ship was the experience of the U. S. S. *Von Steuben,* returning home in June, 1918. A lookout on the *Von Steuben,* a bright young apprentice lad, sighted the wake of a torpedo running toward the ship at a distance of only about 500 yards.

As the speed of a torpedo is over 25 knots, there was no margin for error. The lookout was on the alert and made his prompt report, "Torpedo wake bearing 270 (port bow)." The helm was at once thrown over, and the torpedo passed less than 50 yards ahead of the ship.

Here was a case where three brains acted quickly and in coordination, the lookout, Louis Seltzer, the Captain, Yates Stirling, Jr., and the Helmsman. The slightest mistake on the part of any one of the three would have resulted in the loss of the ship.

SPEED

Of all kinds of protection within the ship against submarine attack, high speed was probably the most effective. A submarine under water has only a moderate speed and must use good judgment and also be attended by good luck to attack successfully a vessel traveling two or three times faster.

High speed also enables a quick maneuver. A ship moving rapidly answers her helm more promptly than when going slowly, and therefore can be turned with greater ease to avoid a submarine or the path of a fired torpedo, revealed by its wake.

Every endeavor was made to assure all transports making their maximum speed while passing through the danger zone. This called for care in organizing convoys, as the speed of the convoy is the speed of the slowest ship.

ZIGZAG TACTICS

Zigzag tactics were introduced by the English. At sea it is a simple problem to observe, and then estimate the course and speed of a ship if both remain steady — otherwise not.

Various methods of zigzagging, that is, making radical changes of course at irregular intervals, were used in the Cruiser and Transport Force. As all ships had to turn together, each separate method was numbered, and the Convoy Commander had only to signal the number, and then change the plan from time to time further to puzzle the submarine.

Each transport carried a zigzag clock carefully set to Greenwich time and placed in a specially screened box in front of the helmsman. This was to assure that all ships put their rudders over simultaneously, on the dot, in order to minimize the danger of collision.

If it had been the practice to follow only one zigzag plan, a submarine might follow in the wake of a ship, note and record each change of course, and then act accordingly — also spreading the news to other submarines.

This was the case of one freighter which was picked off from a slow convoy by a U-boat Captain who trailed until he got the plan, then steamed ahead to a favorable attacking position, and "let go" a torpedo which sunk one of the ships.

TACTICS TO DESTROY

Tactics to destroy, to harass, to make the submarine the hunted one as well as the hunter, were useful, both to lessen the enemy's numerical strength and also to damage his morale. All vessels in the Cruiser and Transport Force carried guns and depth bombs, and were on the alert to use ramming tactics whenever opportunity offered.

Mention has already been made of target practice at sea, and of the non-ricochet type of shell developed to dive and follow an underwater trajectory and explode against the submerged U-boat.

The gun was chiefly useful to compel the submarine to keep under water and use his torpedoes at a disadvantage. It was difficult to hit a periscope and if a lucky hit was made no lasting damage resulted as spare periscopes were carried. Still, the presence of the gun was important, both to embarrass attack, and also to destroy the U-boat when for any reason it was forced to come to the surface.

Submarines are vulnerable, and as a general rule, they did not like to take chances on being hit by gunfire. The policy of arming merchantmen, together with the convoy system, upset the plans of the larger type of U-boat cruisers, because they had no opportunity to attack on the surface, except in the face of an effective gunfire, while their large size made them unhandy in making submerged attack.

Torpedoes, moreover, were expensive and could not be carried in large numbers. On the whole, it may be concluded that the gun was an important factor in defeating the submarine.

DEPTH BOMBS

Depth bombs, variously known as depth charges or water bombs, were dropped over the stern of a ship, or thrown in pairs, simultaneously to a distance on either side of the vessel, by means of a "Y" gun.

These bombs were fitted with a hydrostatic valve, operated by the weight of water, so that the charge — 300 to 600 pounds of TNT — exploded at a certain depth. If not near enough to blow in the U-boat's sides, or to disarrange the delicate internal machinery and fittings, at least it damaged the morale of the crew.

SMOKE SCREENS

Smoke screens to hide the convoy were sometimes made by escorting destroyers, or by smoke boxes thrown overboard, or by smoke funnels mounted on the stem filled with a phosphorous compound which emitted a dense black smoke.

CAMOUFLAGE

Wide use was made of camouflage painting of hulls and exterior fittings of all types of ships, to confuse the enemy in estimating the course, speed and size of his quarry.

For a long time, it was generally thought that camouflage acted like the invisible cloak of the knight in the fairy tale, which of course it didn't.

There were various styles of camouflage just as there were different kinds of zigzag. Some camouflaging was so effective that the course of the ship was disguised as much as 90 degrees. Once an officer of the deck reported that a ship had been sighted heading directly across his bow, when as a matter of fact she was going in the same direction.

Any one living in New York City during the war had opportunity to see from Riverside Drive the various designs of camouflage. Some of these were fantastic, but the majority were known as the "dazzle system," which sufficiently indicates the style.

RADIO

All transports and their escorts were required to confine to a minimum the use of the radio telegraph. A receiving vessel can judge the approximate distance of the transmitting vessel by the strength of the sound. The Germans had also developed their radio direction finders to a high degree of efficiency, so we simply cut out using the radio, except in cases of extreme urgency.

An alert radio "listening-in" watch, for receiving SOS calls and information from destroyers and shore stations, however, was always maintained to enable Group Commanders to lead their convoys so as to give torpedoed vessels and submarines sighted a wide berth.

Submarines frequently sent out SOS calls to attract rescue vessels to their vicinity, but the German radio apparatus produced a sound of distinctive pitch which the trained ears of our operators usually detected,

DARKENING SHIP

One of the most important measures of protection was the complete darkening of the ships at night. All ports and openings through which light might show outside were carefully sealed.

It was with the greatest difficulty that ships were taught that to darken ship was to make them as black as starless night. On the first expedition the strictest orders were enforced from the beginning. Each ship had to report to the flagship every morning what lights she had seen on other ships during the night.

It was not an easy task to make thousands of men who had never seen a ship before, realize they could neither smoke after sundown or even carry matches. It is a fact that the light of a cigarette may be seen for a half mile, an ample radius for exact submarine torpedo practice, hence the importance of absolute darkness.

There were many kicks at first against the seizing of electric torches and matches, but like many other objections, necessity overruled them. "You shall not take my matches," said a Tennessee Mountaineer, as he stepped on the gangplank of a transport. "Just watch me," replied the Naval Master-at-Arms, and immediately passed the trooper's first line of defense.

Major General Lejeune told me, on my after-the-war visit to Germany, that he considered the greatest hardship the troops had to endure was being deprived of smoking on their night marches, and also of the traditional campfires in bivouac. The airplane's eyes were as keen as the "sub's." Ashore as well as afloat, darkness, and a great deal of it, was the order of the night.

WATER-TIGHT INTEGRITY

Water-tight integrity was another point which received careful attention. At all times at sea, water-tight doors were kept closed in order to retain buoyancy in the event of being torpedoed. Water-tight bulk-heads were carefully inspected, and other measures, too numerous to mention, were adopted to guard against the flow of water from an injured compartment into another part of the ship.

I have often thought with satisfaction of the doctrine Captain D. E. Dismukes enforced in the *Mount Vernon,* "Men, remember that one torpedo cannot sink your ship, *but* keep your water-tight doors shut." The epigram suggests the older one, "Trust in God, but keep your powder dry." When the day arrived for the *Mount Vernon,* although badly damaged, she got into port. Her men said, "Of course we are all right, only one torpedo hit us."

The Burney gear was a protection for capital ships against mines, and was invented by a British Naval Officer. It consisted of two otters, designed to tow under water, on a level with the keel, one on either side, at the end of steel cables, at a distance from the ship, and well forward.

Unless the sharp stem of the ship came in direct contact with the mine (something not likely to happen), the mine would slip along the cable to the otter, where the otter's teeth, a kind of shears, would automatically cut the mine adrift, allowing it to float to the surface at a safe distance from the ship's side, where it could be destroyed by gunfire.

Had the Cruiser *San Diego* — sunk by a mine off Long Island — been fitted with Burney gear, she doubtless would have escaped, as the Battleship *New Hampshire* did later on, when, while cruising along our coast, a mine was plucked by one of her otters and then destroyed with gunfire.

Chapter Six - Development of Transport Force— Returning the Army

It was soon evident that now the way was open we would send hundreds of thousands of men to fight in France. The Transport Force grew apace. All available American ships were requisitioned, and, in addition, the War Department arranged with foreign governments for as many ships as could be spared to lend us a hand in getting the soldiers across; England, of course, furnished by far the greatest number, Italy a few, France a few, and Brazil one. We secured three Dutch ships also. To protect these vessels in their ocean voyage, all of the United States cruisers were employed, reenforced by a division of French cruisers, commanded by Rear Admiral Grout. Of the latter the *Dupetit-Thouars,* commanded by Capitaine de Fregate Papue, was torpedoed and sunk while engaged in escorting one of our merchant convoys.

In the early operations of the transports, difficulties were encountered which were inevitable in the rapid development of the Force. The greatest of these was due to inadequate docking space and insufficient lighters, tugs, barges, coaling facilities, railroad transportation and other equipment in the French ports of debarkation. Remarkable results, however, were obtained with the material at hand, and as the organization was perfected and experience obtained, conditions improved.

During the first six months of 1918 the Transport Force increased rapidly in numbers. The speed of operation also continued to improve as the machinery defects were overcome, the coaling difficulties solved, and the organization standardized and consolidated. The delays in the ports of debarkation, St. Nazaire and Brest, were materially reduced as the Army obtained additional labor and equipment for receiving the transports' troops and cargoes.

In January four convoys, averaging three transports to a convoy, were dispatched with 25,662 troops. In February three convoys averaging five ships each were dispatched, carrying 39,977 troops.

The plans made for the increase of troop movement in 1918 developed the necessity for another outlet than New York, in order to reduce port congestion, to improve railroad transportation ashore and to increase facilities for coaling and repairing. Newport News, Va., was agreed upon by the War and Navy Departments as an additional port of embarkation, and sufficient ships were assigned to that port to provide for the carrying of 40,000 troops per month from Newport News to France.

On April 1, 1918, Rear Admiral Hilary P. Jones, Commanding Division Four, of the Cruiser Force, was assigned additional duty as Commander of the Newport News Division of the Transport Force. He established headquarters at Newport News and as my representative in that port proceeded at once to organize and operate the Newport News Division.

The procedure for the convoys was as follows: The troopships were sent over in groups, and these groups, as a rule, were composed of not less than four, or more than twelve ships. Altogether 88 groups sailed from the United States from June 14, 1917, to December 2, 1918. Each group usually started in two sections, sailing simultaneously, one from Hampton Roads, and one from New York, and joining up at a prearranged rendezvous off the coast. They were accompanied to the hundred fathom curve by a cruiser, destroyers, chasers, submarines and aircraft. Then the light craft returned to port and the cruiser continued on to a certain meridian where the convoy was met by the European destroyers and taken through the danger zone. The voyage from the United States to France averaged twelve days, except for the fastest ships. The *Leviathan, Northern Pacific,* and *Great Northern,* usually sailed together and without escort to the overseas rendezvous, their high speed being their best protection.

As the need for rapid transatlantic troop transportation became more pressing, every effort was made to increase the troop carrying capacity of the individual vessels to the maximum that was considered safe. Careful calculation of all available space was made and additional bunks installed. The increase was made during the time of lay-over in American ports and in no cases was the sailing of a transport delayed by this work.

The great German drive in March, 1918, produced an urgent and imperative call for more troops. Notwithstanding the fact that the American ships were carrying many more troops per ton than the foreign ships, an increase of 40 per cent to 50 per cent was obtained in some of the larger ships by the "turn in and out" method; that is to say, the extra men carried took turns with others in sleeping in the bunks. In other words, the bunks were always occupied. This was carried out only in the fastest ships, where the discomfort lasted for the shortest time, and the high speed of the ship rendered them fairly immune from torpedo attack. The troop capacity of the *Leviathan* was thus increased 100 per cent from 7,000 to 14,000.

Coaling and repairs were always pushed at top speed, working 24 hours of the day. At one time in New York harbor the coaling became a serious proposition, owing to the unsatisfactory condition of labor and the severe weather, and it looked as if the ships would be held up; this was just at the time when the troops were most needed. But a crisis was avoided by commandeering the coaling equipment, and carrying on with our own people.

Until May, 1918, almost all of our troops were embarked in our own Naval transports; but after that date the call for more men became so urgent that the great British liners were called in to assist. All hands had to pull together to defeat the German armies which were overrunning France. It was a case of the Allies ' domination or downfall. As many of the British ships had been taking over Canadian troops, they were ready to receive and transport our soldiers. From first to last 196 British vessels were employed in this work.

On July 1st, a year after the operation began, the total number of troops in France and embarked for France, was 1,029,003; of these 456,854 had been sent over in British ships; 524,457 in American ships, 18,476 in French and Italian ships, and 29,218 in Italian ships leased by the British government. On June 5, 1918, I had the gratification of addressing the below quoted commendatory letter to the personnel of the Cruiser and Transport Force upon the completion of the first year of service:

"At the end of our first year of service as the Cruiser and Transport Force, I desire to congratulate the Flag Officers, Captains, officers and enlisted men on the excellent work they have accomplished, and to express my personal as well as official appreciation of their splendid loyalty and cooperation in all the exacting, arduous and hazardous duties that have been assigned to us.

"The preparation in three months of the fleet of ex-German ships, which for three years were idle, and worse, at their piers, was in itself a great achievement.

"The organization, supply and sanitation of types of ships, entirely new to the Navy, for a service overseas of the most vital importance, not only to this Country but to our Allies, presented serious and complex problems, which have all been happily solved by' your intelligence, zeal and ability.

"The safe conduct of transports ladened with troops through seas infested with submarines has won universal commendation. The loss of only two transports in the transportation of hundreds of thousands of troops testifies to the skill, courage, and seamanship of the Commanders; and in the two cases of loss the highest and best traditions of the service were maintained, speaking volumes for their organization and discipline.

"I wish to take this opportunity of impressing upon all Captains under my command, that in every position of stress and trial which may come to them, I am confident of their ability and judgment to meet the situation most creditably; and whatever happens they may always feel sure of my sympathy and support."

From July, 1918, until the signing of the Armistice, the troops crossed at the rate of nearly 10,000 per day. In July, 1918, 311,359 were transported in shipping of all kinds. Of this number 56 ½ per cent, or 175,526, were carried in British ships. This was the greatest number transported in any one month under the British flag. We carried only 36 per cent the same month, and this fact probably gave rise to the then prevalent but erroneous belief that American ships were carrying only about one third of the troops.

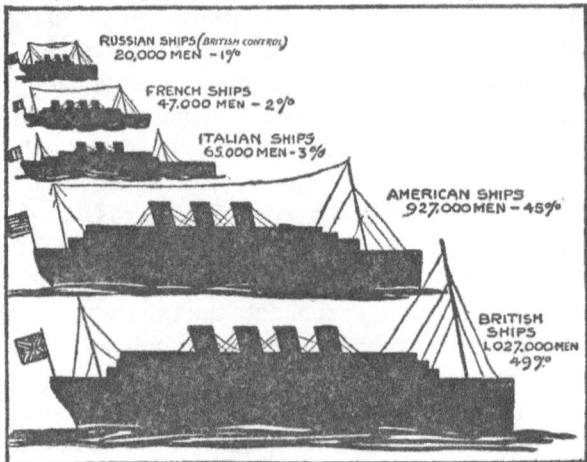

American troops carried by ships of each nation

The actual operation of our transports continued to increase in efficiency up to the signing of the Armistice. Additional destroyers having been sent abroad for escort duty, it became possible to sail medium speed (13 to 14 knots) troop transport convoys from New York at 7 day intervals and fast troop transport convoys 15 ½ knots and above) at 5 day intervals.

On November 11, 1918, the Armistice was signed and the war activities of the Force were ended. Up to the signing of the Armistice a total of 2,079,880 of the A. E. F. had been transported in 1142 troopship sailings. This number was carried as follows:

	Total	Percentage of total carried
By U. S. Navy Transports..................	911,047	43.75
By British Ships...........................	1,006,987	48.25
By British leased Italian Ships..............	68,246	3.00
By other U. S. Ships.......................	41,534	2.50
By other foreign ships, French, Italian, etc.....	52,066	2.50

Note: Total carried in United States ships was 952,581; percentage of grand total, 46.25.

RETURNING THE ARMY

With the signing of the Armistice the Eastward flow of troops ceased and the return movement began, at first slowly. Transports continued sailing on a slow schedule without escort and not in convoy. Advantage was taken of this comparatively inactive period to give certain vessels, including the *Leviathan,* a much needed overhaul.

Soon the public began to demand the speedy return of the overseas Army so that the civilian army could be demobilized. As was to be expected, the British and other foreign ships which had carried a little more than half of our soldiers to France were rapidly withdrawn from this service and most of the work of repatriating this Army of two million fell to the lot of the Cruiser and Transport Force.

When submarine activities ceased, relieving the necessity of numerous anti-submarine precautionary measures incompatible with crowding beyond certain limits, it was possible to increase the troop carrying capacity of vessels in use at that time. The following are examples of this work, which was at once proceeded with in all vessels:

Leviathan Increased from 10,000 to 12,000
Agamemnon " " 3,000 to 5,500
America " " 4,900 to 7,000
Geo. Washington " " 5,500 to 6,700
Orizaba " " 3,100 to 3,900
Siboney " " 3,100 to 3,900

This work was carried on as opportunity offered and without delays to the movements of the vessels concerned. The work was laid out by a joint Army-Navy Board and involved the installation of standees, increase of ventilation, washroom and galley facilities, and life saving equipment. The increased capacity resulted in a very material saving. The troops were necessarily crowded and deck spaces for airing and exercising troops limited, but in no case was this overdone and no justified or serious complaint was received.

The force continued to expand and 56 cargo vessels were converted by the Army into troop transports and added to the Force. The majority of the officers on these converted vessels were enrolled in the Naval Reserve and continued in their same position when the ships were commissioned. On a few of the larger vessels it was considered advisable in the interests of efficient organization and administration to place regular officers in command until the Reserve Officers had been indoctrinated with the methods of the Navy and of the Transport Force. These reserve officers quickly absorbed the spirit of the Navy and the mission they had to accomplish, and are deserving of the highest praise for their excellent work and devotion to duty.

In December the battleships of Force Two and the armored cruisers were assigned for the transportation of troops. The battleships carried an average of 1,100 troops, and the armored cruisers about 1,750. The former operated on a forty-day round trip schedule and the armored cruisers on a thirty-day schedule. When the fleet was reorganized in the Summer of 1919 orders were received to withdraw battleships and cruisers from troop transportation service.

On April 19, 1919, the *Kaiserin Auguste Victoria,* the first of the nine German vessels allocated by the Peace Conference to help return the U. S. Army, arrived in New York for conversion into a troop transport. These vessels were converted by the Navy on an average of less than 14 days per ship and at a cost per troop of about $40.00 for material, labor and overhead charges. Under the Army the average time in port of the cargo vessels was 75 days and the cost per troop for conversion was about $78.00. This comparison is made not as a criticism but to emphasize the obvious lesson the war has taught that naval handling and operation of troopships makes for economy and efficiency. The results obtained by the Navy were due to intelligent planning and supervision, born of knowledge of ships and experience gained from previous work of this nature. The transportation of troops had developed into a science and methods had been revolutionized. The German vessels were converted to carry a total of 3,997 officers and 39,132 men. The giant *Imperator* was fitted out to carry 9,000 troops and 1,400 first class passengers in a period of 10 days.

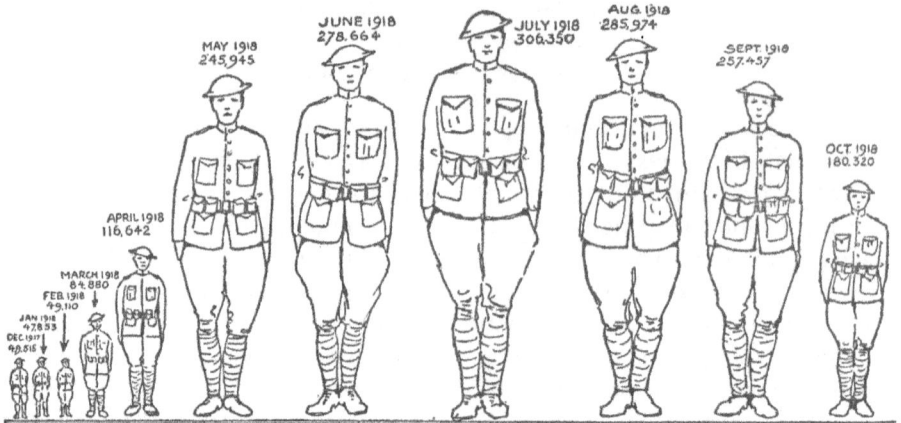

NUMBER OF MEN TRANSPORTED MONTHLY TO FRANCE

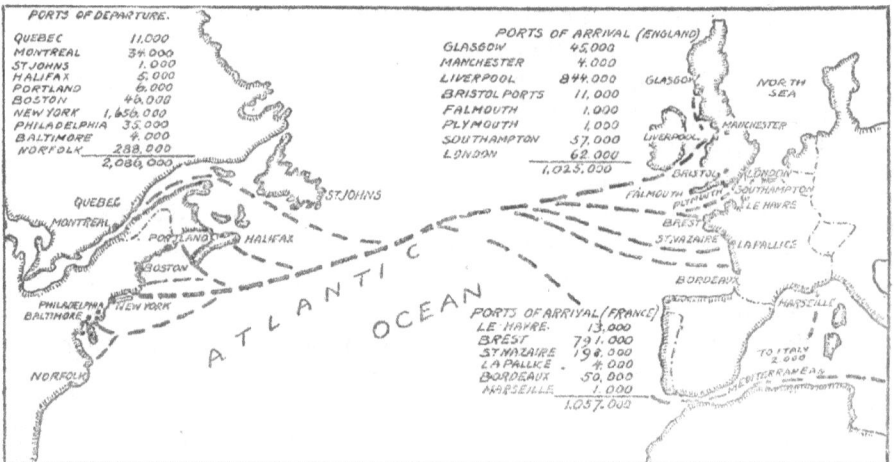

PORTS OF EMBARKATION IN AMERICA AND DEBARKATION CENTERS IN EUROPE

Beginning January 1st, the troop movements gradually increased and the number of troops carried by this Force increased from month to month until the maximum was reached in June, when the total of 314,167 were actually landed in the United States. This exceeded the maximum carried overseas by all U. S. and Allied vessels in any one month during the war. When the troop movement reached its highest efficiency, the average cycle of troop transports was 25 days, and of the converted cargo vessels about 35 days. For certain 10-day; periods, the average cycle of the former reached the low level of 21 days, and the latter 29 days. It was anticipated that the troop transports required approximately a 30-day cycle, and the cargo vessels a 40-day cycle, but the increased efficiency of loading troops in France, and of repair, provisioning and coaling, enabled us to exceed the estimated speed of repatriation of troops.

The maximum number of vessels assigned to and operated by the Force for the transportation of troops was 142, with facilities for carrying 13,914 officers and 349,770 men.

The following table gives the total monthly arrivals in United States ports and number of passengers carried from. January to June, 1919:

	Vessels	Eastbound	Westbound
Jan.	47	....	97,039
Feb.	41	....	96,368
Mar.	67	....	165,312
Apr.	87	....	243,397
May	108	....	278,600
June	115	....	314,167

Of the above westbound passengers New York handled 778,318; Newport News 330,398, and other ports 141,389.

Until April, 1919, practically all activities in home ports were confined to New York and Newport News. At this time the Department directed that the District Supervisors, Naval Overseas Transportation Service of the 1st, 4th, and 6th Naval Districts, be my representatives in Boston, Philadelphia, and Charleston respectively. These officers performed their functions most efficiently and vessels landing at these ports were prepared for sailing with a minimum delay. The following table shows the troop movement activities in the three ports mentioned for April, May and June, 1919:

	Vessels	Troops
Boston	23	66,091
Philadelphia	20	41,141
Charleston	14	34,157

MATERIAL

The material conditions of ships in the Cruiser and Transport Force were as a whole on a very high plane, when the very severe operating conditions are considered. The cruisers stood up very well indeed for two years of most exacting duty, and hard steaming. With the exception of the U. S. S. South Dakota, which broke a propeller shaft, all cruisers maintained their schedules throughout the war and while in use as troop transports. These vessels averaged about one month Navy Yard overhaul for the two years. The troop transports were more easily maintained, due to more rugged and simpler machinery installations.

During the year ending July 1, 1919, the following transports had extended overhaul or repair periods:

Aeolus Boiler...engines, auxiliaries.

Agamemnon ...." " "
Great Northern...Turbines.
Harrisburg..Condenser tube sheets — boilers.
Kroonland..Main engines — auxiliaries.
Leviathan...Turbines.
Mallory..Engine foundations.
Henderson.." "
Sierra..." " — crank shafting.
Powhatan ....Boilers, engines, auxiliaries.
Pocahontas..." " "
Von Steuben...Boilers.
Mount Vernon..Repairing torpedo damage.
America..Sinking at dock.
K. der Nederlanden...Boilers.

The *Tenadores* was lost by grounding in the fog off St. Nazaire, on December 30, 1918; the *Northern Pacific* went aground near Fire Island Light on January 1, 1919, but was later floated, towed into port and repaired; and on January 11, 1919, the *Graf Waldersee* was in collision and beached, but was also floated and the comparatively slight damage done was repaired: no lives were lost in these casualties.

Other transports maintained their schedules with very little, if any, delay. Generally speaking, all vessels decreased the amount of assistance required from outside sources and the volume of repairs per unit was materially decreased during the year. This was due to improved organization and training of personnel and to a generally improved material condition, as a result of superior methods of maintenance employed in Naval practice.

On September 1st, I was relieved as Commander of the Transport Force and having been promoted to the rank of Admiral hoisted my flag in the *South Dakota* as Commander-in-Chief of the U. S. Asiatic Fleet and station. Nearly all the troops having been returned, the transport fleet was rapidly demobilized under the direction of my successor, Rear Admiral C. B. Morgan.

During September and October, 42 transports were turned over to the Shipping Board for further transfer to owners, while 15 were turned over direct to the Army Transport Service. On October 31, 1919, only 3 vessels, the *George Washington, Martha Washington,* and *Pocahontas,* were retained under Naval operation; these were transferred to the supervision of the Commandant of the 3rd Naval District and the Transport Force was finally demobilized and disbanded. While this is true of the ships, there is still a link which binds the personnel.

Shortly after the Armistice, Commander Robert Henderson suggested that the spirit of comradeship and service developed during the war be perpetuated by a "War Society of the Cruiser and Transport Force." This suggestion was received with enthusiasm. The Society was formed and a constitution with by-laws was drawn and approved. In due time Lieutenant De C. Fales was directed to incorporate the Society under the laws of the State of New York; Ensign R. B. Lanier was elected Treasurer, and Lieutenant Clifford N. Carver, Secretary.   These officers, all of whom performed excellent war service in the force under my command, have ably managed the affairs of the Society. It has expanded rapidly and is fast establishing itself as one of our national institutions.

The following tables will be found in the Appendix:

Table A   Organization of Cruiser and Transport Force July 1, 1918.

Table B   Report by months of Transport and Escort Duty performed by United States and Foreign Navies up to signing of Armistice.

Table C   Report by months of transport duty performed by U. S. Navy and all other ships, United States and foreign, in returning troops and other passengers to United States prior to signing of Armistice.

Table D   Report by months of transport duty performed by U. S. Navy and all other ships, United States and foreign, in returning troops and other passengers to United States since signing of Armistice.

Table E   Complete list of all U. S. Naval Transports and U. S. Battleships and Cruisers engaged in transporting troops to and from France between the dates of June 14, 1917, and October 1, 1919, which were operated under the Command of the Commander of the Cruiser and Transport Force.

Table F   Sick and wounded returned by the Cruiser and Transport Force.

Table G   Record of ten leading troop-carrying ships.

Chapter Seven - Sinking of Antilles— Finland Torpedoed

The *Antilles* arrived in. Brest, France, from New York, on October 7, 1917, with approximately 1,100 troops and officers. On October 15th she sailed for the United States in convoy with the U. S. Naval Transport *Henderson,* and the Army Cargo Transport *Willehad,* escorted by the U. S. S. *Alcedo,* U. S. S. *Corsair,* and U. S. S. *Kanawha.*

The *Antilles* was an Army transport manned by merchant officers and crew, and carrying an additional detail from the Navy of two officers, two gun crews, quartermasters, signalmen and wireless operators. The senior Naval officer was Lieutenant Commander D. T. Ghent, U. S. Navy.

On the second night out of Brest the weather was intermittently squally and foggy, with a fresh easterly breeze and rough sea. During the evening the increasing sea forced the *Kanawha* to change course and leave the convoy. Early in the morning of the 17th the fog had cleared, permitting a view all around the horizon.

At 6:48 A. M., while in Latitude 48° 10' North, Longitude 11° 20' West, the quartermaster of the watch sighted a torpedo headed for the ship from two points abaft the port beam and about 400 yards distant. The torpedo was sighted almost simultaneously by the officer of the watch and the signalman. The rudder was immediately put over to turn the ship to starboard in order to parallel the course of the torpedo and reduce the target area presented by the full length of the ship.

There was not sufficient time, however, and within half a minute after it was sighted the torpedo struck the port side of the ship and exploded in the after part of the engine room. The effect of the explosion was terrific; the ship shivered from stem to stern, and almost immediately took a heavy list to port.

One of the lookouts in the main top, although behind a canvas screen reaching to his shoulders, was whipped out of the top, thrown to the deck and instantly killed. The guns were manned at once by their crews, who searched the surface of the water for a glimpse of the submarine, but not even a periscope was sighted, nor was anything ever seen of the submarine.

The explosion of the torpedo completely disabled the engines and wrecked the engine room, which was flooded almost instantly, and within a few moments the fire room and a cargo hold just abaft the engine room were also flooded.

The ice machine in the engine room was wrecked and the escaping fumes of ammonia overcame the engineers who had not been killed outright by the explosion or thrown into the moving machinery. Of the engine room crew, only one man escaped; he was an oiler who happened to be on an upper grating at the time and succeeded in climbing up the hatch. All of the fire room crew were killed except two men who climbed to the deck through a fire room ventilator.

Lieutenant Commander Ghent, seeing that there was no chance for the ship to remain afloat, gave the order to abandon ship shortly after the torpedo struck.

Navy radio electrician C. L. Ausburne went to his station in the wireless room, relieving the operator on watch, and commenced sending out the call for help and the ship's latitude and longitude. Ausburne remained at his station, going down with the ship, and in reporting his act to the Navy Department, I wrote as follows:

"At the time the *Antilles* was torpedoed, Ausburne went to his emergency station at the radio key in the Wireless Room. It was his duty to send the 'SOS' distress signal and he evidently sacrificed his life in persistent endeavor to accomplish this duty. For this service, in which he distinguished himself conspicuously by gallantry and intrepidity at the risk of his life above and beyond the call of duty, I recommend that a posthumous Medal of Honor be awarded and sent to his next of kin."

The boat falls were manned without confusion, and the boats lowered with considerable difficulty. The ship listed to port and began to settle by the stem, making it impossible to lower two of the boats into the water. One of them had been destroyed by the explosion and. the boat davit of another had been damaged so that the boat could not be swung out over the water. The heavy seas swamped two boats alongside and only four of them got clear of the ship. Life rafts were launched and the men who could not go in the boats jumped into the sea with their life belts on and swam to the rafts. The temperature of the water was 53° F.

The ship was seen to be sinking rapidly, and the forward gun crews, who were still standing by their guns under command of Lt. (j.g.) R. D. Tisdale, U. S. Navy, were ordered to leave their guns and get clear of the ship. The after guns at this time were submerged.

Ghent, engaged in seeing all hands clear of the ship, was walking aft to order some men in the water alongside to swim away to escape the suction when he himself was picked up by a heavy sea breaking over the deck and washed overboard into a tangle of floating wreckage.

At this moment the bow of the ship rose vertically in the air and she began to slip rapidly, stern first, into the sea. The smokestack was just above Ghent's head and about to carry him under when the explosion of the boilers produced an upheaval of water which washed the life raft to which he was clinging a few feet clear of the stack.

The ship disappeared into the sea only six and one-half minutes after she had been torpedoed. Sixteen enlisted men of the Army, returning to the United States, four of the Navy, forty-five of the merchant crew, one civilian ambulance driver and one colored stevedore, were lost, making a total of sixty-seven out of 234 persons on board.

Most of these casualties were probably victims of the explosion.

When the *Antilles* was torpedoed the *Henderson* and *Willehad* turned to starboard and port respectively, and proceeded at full speed. The yachts *Alcedo* and *Corsair* returned to the *Antilles* and circled about her on lookout for the submarine, one vessel patrolling while the other rescued the survivors.

The ship was abandoned in excellent order and without undue excitement. The saving of 71 per cent of those on board in the rough sea that was running, while the ship went down in the unusually short time of six and one-half minutes, was a creditable performance.

The gun crews, in particular, displayed coolness and daring, remaining quietly at their guns and searching for the submarine while the ship was sinking, hoping that they might get in one shot at least.

Later, one of the gun crew, unable to find a raft, swam to a large ammunition chest which was floating about upright and perching himself upon it, calmly waited to be picked up. When the Corsair bore down directly for him he signaled to her in semaphore — "Keep clear, this box contains live ammunition!"

The following is excerpted from a letter written to me by Captain L. W. Steele, Jr., U. S. N., then commanding the U. S. S. Henderson, next ship to the ill-fated Antilles.

U. S. S. Henderson. October 18, 1917.

My dear Admiral,

Yesterday we witnessed the sinking of the poor old *Antilles,* our companion of all three voyages. She was struck at 6:47 A. M. Greenwich Mean Time, and as we were in longitude 11° 22' W., this time was just about sunrise.

We were in column, this ship leading, followed by Antilles and a freight steamer named *Willehad,* or something like that. Our escort, the yachts *Corsair* and *Alcedo,* were some distance, 3,000 yards, ahead of the column. We were zigzagging, plan one, and you can sketch the position we were in at 6:47. Our speed was 10 knots. The sea was a bit choppy, with enough white caps to make discovery of a periscope extremely difficult.

I was attracted by the sound of *Antilles's* whistle, and looked around and saw the explosion of the torpedo against her port quarter, about opposite the well deck. The water rose as high as the hounds of her mast. This was followed almost immediately by an internal explosion aft, and she began to list to port. I remember Ghent's telling me that all her heavy machinery, stores, etc., were on the port side.

In the meantime I was busy taking *Henderson* away from that vicinity. It is not a pleasant feeling, Admiral, to run away from a ship in such a predicament, and it should be strongly emphasized in orders so that a person doing it will not feel such a deserter. But the yachts were already returning.

The next glimpse I had of *Antilles* she had turned head on to our position, and all her boats seemed to be lowered to the water. And then, in a very short time, I looked again, and there she stood, upright against the red morning sky, looking like some strange monster. She sank vertically and rapidly, but silently, and *Antilles* was no more.

Four, it was, I think, of her boats we counted, and the water dotted with heads. The water closed over her at 6:53 ½ — it had taken six and one-half minutes! I do not like to picture the awful confusion caused by her standing vertical in the water — what crashings there must have been! We hope and pray that many of those men were picked up, but there must have been many casualties. It made many of us very quiet and thoughtful yesterday — some did not eat a bite all day after seeing it.

It is practically certain that the sub which sank *Antilles* fired at *Henderson* and missed. There was no reason to pass us by for a smaller ship half as well armed.

Finland torpedoed

The Transport *Finland* arrived in France on October 7th in the same convoy with the ill-fated *Antilles,* and sailed again for the United States in the early morning of October 28th in company with the cargo vessels *Buford* and *City of Savannah.* The escort was made up of the armed yachts *Alcedo, Corsair* and *Wakiva,* and the destroyers *Smith, Lamson, Preston* and *Flusser.* The speed of the convoy was eleven knots. Commander S. V. Graham, U. S. Navy, was the senior Naval Officer on board the *Finland.*

At 9:27 of the same morning of departure, while in Latitude 46° 49' North, Longitude 6° 21' West, a torpedo fired at the *Finland* was sighted about thirty degrees abaft the starboard beam at a distance of about 200 yards. A few seconds later it struck the starboard side under the bridge before the ship could be maneuvered to avoid it.

Both the *Finland* and the *Antilles* were manned by civilian crews, the Naval personnel on board being additional to safeguard the ship against the enemy and to take charge in emergency. The Finland was carrying home the survivors of the *Antilles,* and the majority of these merchant sailors were a very low class of foreigners of all nationalities, the sweepings of the docks, shipped just before sailing from New York for one voyage only. The terror from which the men of the *Antilles* had not yet recovered had been communicated by their stories to the crew of the Finland, which was made up of the same type of men.

The result was that when the torpedo struck the ship both the crew of the *Finland* and the survivors of the *Antilles* rushed to the boats and began lowering them. Some of the boats were in the water and some were capsized before the Naval officers and ship 's officers gained control of the situation.

At the same time the engine room and fire room crews left their stations and rushed on deck, which was contrary to orders. These men were finally driven below with the aid of a revolver and a heavy wooden mallet, and the engineers' stations were again manned.

As the ship began to list heavily to starboard, the other boats were lowered in a more orderly manner, with the passengers and some of the crew in them.

The damage was found to be confined to one of the cargo holds, which was flooded; the engine room and fire room compartments were intact. The list which the ship had taken did not increase, and Graham decided that the ship would remain afloat and that she could be worked under her own steam.

A number of men who were drifting about in the boats were taken on board and the remainder left to be picked up by the *Wakiva* and *Alcedo,* which were standing by and rescuing men in the water. At 10:45, the Finland shaped a course for Brest and anchored in the harbor the next morning.

As a result of this experience with the ignorant and unreliable men composing the crew of the *Finland,* the Court of Inquiry which investigated the circumstances, recommended that all troop transports be officered and manned entirely by Navy personnel. This reinforced the recommendations I had previously made and was done as rapidly as possible.

Chapter Eight - Loss of President Lincoln— Covington Torpedoed and Sunk

On the 29th of May, 1918, a convoy consisting of the Troopships *Rijndam, President Lincoln, Susquehanna* and *Antigone,* escorted by destroyers, sailed from Brest, France, on the return voyage to America.

At about sundown the next day, having almost passed through the so-called danger zone of supposed greatest submarine activity which would be completely cleared by the next morning, the destroyer escort left the convoy to make rendezvous with and act as escort for another convoy carrying troops eastward bound. It was our policy always to provide maximum escort strength for ships loaded with troops. The homeward bound ships had to run chances when the exigencies of war required that chances be taken.

At this time the German submarine *U-90*, Captain Remy in command, was on her cruising station about 300 miles to the westward of the French Coast. She was making five knots on the surface, when, about one hour after midnight, her lookout sighted in the moonlight a convoy, distant about 2,000 yards.

Captain Remy, at a safe distance, trailed this convoy, which included the *President Lincoln,* and finding that he had superior speed, he made a wide detour on the surface in the hope of getting ahead m position to attack the next morning.

In this he was successful, and being in a favorable position, he submerged before the convoy was near enough to sight the smaller submarine. Remy singled out the President Lincoln for attack as she was the largest in the group. The submarine approached from the port bow, intending to close just ahead of the lefthand ship, the *Rijndam,* her quarry being the second ship, while the *Antigone* and *Susquehanna* were the third and fourth ships from the left of the line and, therefore, on the other side of the *Lincoln.*

The submarine Captain was skillful in his maneuvering, except that he got a little nearer the *Rijndam* than was comfortable for him, and narrowly escaped being rammed by that vessel, as is shown by the accompanying sketch. No doubt he was confused by the zigzag courses the convoy was steering.

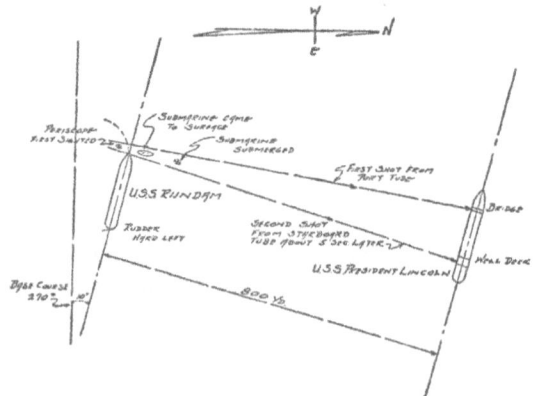

Torpedoing of the President Lincoln

The weather conditions were favorable for the submarine in that they made it difficult for the lookouts to detect the periscope. The wind was southeasterly and stirred up numerous white caps on which the sunlight glittered, making it practically impossible to distinguish a periscope and its wake at any great distance.

It was not until the firing exposure of the periscope, made almost directly under the port bow of the *Rijndam,* that the transport lookouts saw and reported the enemy. Captain Remy, however, had already fired a salvo of two torpedoes, closely followed by a third, and his aim was good. Even his passing so close to the bow of the *Rijndam,* since he missed being rammed, worked in his favor because the guns of that vessel could not be brought to bear upon his periscope before he totally submerged.

The first two torpedoes fired were running close together and one of them was near the surface, almost broaching; these were heading for the forward part of the *Lincoln.* The third one was a little behind the other two, and headed toward the after part of the ship. The lookouts on board the Lincoln sighted the torpedo wakes heading for their ship and it looked to them as though they had been fired by the *Rijndam.*

Lieutenant Wesley G. Martin, U. S. N. R. F., Officer of the Deck, immediately had the helm thrown over and cut in the general alarm switch. A few seconds later, before the ship could answer her rudder, two torpedoes hit simultaneously directly under the bridge, throwing up a great volume of water, which drenched every one in the port wing. By this time Captain P. W. Foote, U. S. Navy, was on the bridge and took command of the situation.

Immediately after the first explosion, the third torpedo struck aft, about 120 feet from the ship's stem. At first Captain Foote was in hopes that the ship might be saved, but in about five minutes she was seen to be settling rapidly, and Lieutenant Edward Baker, U. S. N. E. F., Officer of the Watch in the engine room, reported that the after engine room bulkhead had given away and the engine room was flooding. Water was seen rising in hold No. 3, which was just forward of the bridge, and it was realized that the ship was doomed. About twelve minutes after the explosion Captain Foote gave the order to abandon ship.

Boats and rafts were lowered into the water in an orderly and seamanlike manner. The sick were placed in the emergency life saving suits and made comfortable in their assigned boats, the Medical Department efficiently performing their emergency duties under the direction of Surgeon Whiteside.

Surgeon Whiteside and Assistant Paymaster Mowat were last seen standing on the after end of the port side of "C" deck just before the ship sank. Both these officers were lost and it is probable that they became entangled in wreckage and were dragged down by the ship. It is thought that Assistant Paymaster Johnson, who was last seen supervising the launching of rafts and directing his men over the side, suffered the same fate.

Under the direction of the Captain and Executive Officer, Lieutenant Commander Lind, boats and rafts were lowered promptly and without mishap. The men, all wearing life preservers, then slid down the life ropes into the water and were picked up by the boats and rafts. The rafts were tied up to the boats and pulled clear of the sinking ship.

At about 9:20 the Chief Master-at-Arms reported to Captain Foote that the decks were clear of people, and as the ship went down the Chief Master-at-Arms, Executive Officer, and Captain, the latter being the last to leave the vessel, went over the side and swam out to the boats standing by to receive them. The *President Lincoln* kept on an even keel and sank with her colors flying about twenty-five minutes after the first explosion.

Immediately after the torpedoing, the accompanying transports scattered in accordance with the rigid orders requiring them to do so, but they sent messages to destroyers to go to the assistance of the *Lincoln* survivors. The U-boat waited in the vicinity, in the hopes that one of the transports might come back and fall victim to attack. But disappointed in this, the submarine finally came to the surface and steamed toward the boats and rafts which by this time had been secured together.

G. A. Anderson, Seaman 2nd Class, was ordered from a raft to come on board the submarine. Anderson reported that the officers of the submarine treated him very nicely, took him below decks, gave him some cognac and coffee. The U-boat Captain asked Anderson in English where the Captain of the *President Lincoln* was, to which he replied that he did not know, but thought that he had gone down with the ship.

The boat commanded by Lieutenant Commander A. B. Randall, U. S. N. R. F., who had been a passenger on board the *Lincoln,* was then ordered alongside. Ensign C. R. Black, U. S. N. R. F., was in the boat, and Captain Remy of the submarine recognized him, as they had attended the same college in the United States, and called out in perfect English, "We don't want you, Black."

Although the boat containing Captain Foote was closely scrutinized, he escaped detection by removing his blouse and cap and disguising himself as a sailor. Lieutenant Isaacs, however, was taken prisoner and the submarine sailed away and was seen no more. (The adventures of Lieutenant Isaacs will be told in a later chapter.)

Under the direction of Captain Foote, boats and rafts were assembled together before sundown, in one long line, and all hands settled down to make the best of their situation. Signal rockets were sent up every ten minutes and disclosed the men crowded in the boats and roosting on the rafts, laughing and talking and keeping themselves cheered up. Each rocket was the occasion of an outburst of enthusiasm, and the songs most often sung were, "Hail! Hail! the Gang's All Here!" and "Where Do We Go From Here, Boys?"

Shortly after 11:00 P. M. an answering signal was seen in the distance, announcing the approach of help. Soon after the American destroyer *Warrington,* Lieutenant Commander George W. Kenyon, U. S. N., arrived, and about an hour later the *Smith,* Lieutenant Commander Kline, joined in the rescue work. This was a skillful piece of navigation and had there been delay there might have been serious loss of life, especially among the 200 men who were on the rafts. The survivors were quickly taken on board and the destroyers headed for Brest, where they arrived the next day.

Subsequent musters showed that out of 715 souls on board, 4 officers (including Lieutenant Isaacs taken prisoner) and 23 enlisted men, all belonging to the ship's company, were lost. All the Army passengers on board were saved.

In the case of the *President Lincoln,* as in all the casualties suffered in the Cruiser and Transport Force, whether due to the enemy, collision, fire, or other cause, the loss of life was astonishingly small. This was due to the high state of discipline which prevailed, and to the methods and drills previously devised and carried out. Captain Foote in his report states that Lieutenant Commander W. L. Lind, U. S. Navy, the Executive Officer of the ship, was particularly responsible for conducting these drills and also that he rendered valuable service both before and after the *President Lincoln* was sunk.

The Executive Officer is the second in command and by Navy regulations the Captain's representative as the organizer and administrator of the ship. The conditions under which transports were operated in the submarine zone compelled the Captain practically to live on the bridge. His time was taken up with the safe navigation, of the ship through the submarine zone. He was the outside member of the firm; the Executive Officer was the inside member. Details of organization, administration and inspection were necessarily left largely to him; much more so than the Regulations ever contemplated.

One of the novelties of this war was that our very small deep sea merchant marine made it necessary to use regular Naval Officers in manning the seized German ships. It speaks well for their capacity that they were so successful in handling a new type of ship under the trying conditions imposed by troop transportation through submarine waters. Much credit should be given to the Executive Officers of the transports. Their job was complicated and difficult, because they really had two organizations to handle in cooperation, one for the troops and one for the ship.

Lieutenant Colonel W. H. Clopton Jr., of the Tank Corps, U. S. A., was the Senior Army Officer passenger, and the following is quoted from his official report:

I cannot close this report without testifying to the splendid manner in which Captain Foote, his officers and men, conducted themselves from the moment the torpedo hit the ship, until we were picked up by the *Warrington* and the *Smith.*

Confusion, but that orderly confusion which bespeaks of discipline and a thorough understanding of the individual duty and obligation, existed. Life rafts were rapidly pushed overboard. Crews assembled at their stations and all made ready to abandon ship. Cheerfulness prevailed and a hearty response to duty that should make any Conunanding Officer proud of his men.

Captain Foote's subsequent action after the ship had sunk, in assembling all rafts and life boats and giving instructions for the night duties, unquestionably prevented loss of life, through drifting, and expedited the work of the relief ships.

The work of locating us and the reception accorded the survivors by the U. S. S. *Warrington* and *Smith* cannot pass without a word of gratitude. The prompt appearance of these two ships was indeed cheering to the men 440 miles from shore, and the hot coffee, lunch and dry clothes which were given the men were most welcome.

On behalf of the military passengers I desire to express our heartiest gratitude for the manner in which the naval officers and men handled the situation from beginning to end.

(Signed), Wm. H. Clopton, Jr.
Lieut. Col., Tank Corps, U. S. A.

The survivors were transported from Brest to New York on board the U. S. S. *Great Northern,* and upon arrival, Captain Foote submitted the following request:

"The Commanding Officer is glad to report that the officers and men surviving the *President Lincoln* still form practically a complete ship's organization due to the small though regrettable number of its members lost in the engagement with a Gennan submarine. They are still filled with a courageous spirit, and all that is desired is time and facilities to obtain new outfits of clothing, etc., and that they be held together in one organization and assigned to duty in another ship and that they may be continued in the work which they have been performing in the past ten months in connection with transporting our troops to Europe."

It gave me much pleasure to approve this request in the below endorsement:

"I approve the suggestion that, if possible, the survivors of the *President Lincoln* be transferred together to another ship — it is possible that the Department may have in view the commissioning of some new transport in the near future — at all events it is recommended that Commander Foote be assigned to command another ship in my Force at the first opportunity."

TORPEDOED ON THE U. S. S. *President Lincoln*

The following story was told to me by Chief Yeoman Leonard McCallum:

At about 8:55 on the morning of May 31, 1918, while down in the small stores compartment of the U. S. S. *President Lincoln,* I suddenly heard two loud explosions forward, followed a second later by an explosion aft which seemed to shake the whole ship and shove her back in the water. Then the General Alarm sounded and with the roar of our forward guns I realized that we had been torpedoed. Seizing a life preserver I hurried up on deck.

As I reached the main deck, the ship listed slightly toward the port side and the water poured in from that direction, the torpedoes having hit on the port side. However, there was no confusion, every man knew where to go and was hurrying quickly and silently to his station. Joining my division on the port side of "C" deck, I reported to Paymaster Mowat and was ordered to go below and secure the payrolls and cash book. The men of the division were lined up in two rows facing outboard awaiting the order to "Abandon Ship" and watching the accompanying transports tearing away from us. The life boats were being lowered, all orders being shouted from the bridge.

When I returned from the pay office, the order to abandon ship had been given and the men were all working together sliding life rafts over the side. The discipline was perfect, not one man attempting to leave the ship until the order was given. The first rafts to get away from the ship had firemen on them and as they sat on their rafts they sent up cheer after cheer for the forward gun crews who were firing in water up to their knees.

When it came my turn to abandon ship I slid down a line and sort of stepped onto a raft upon which there were five other lads. We started to paddle away from her side but were forced back against her twice. Officers were singing out for all to get away as quickly as possible on account of the suction. That didn't bother the crew much because as some of them tried to paddle away, they'd yell, "Liberty party shoving off, etc." It all seemed more like a picnic.

When we felt that we were a safe distance away from the ship we turned to take a last look at her and what a fascinating sight she was. Our flag was flying, the gun crews were firing, the steam was hissing and above it all, sounding like the death cries of some big old animal, could be heard the mournful shriek of the siren. She seemed to be sinking very slowly, when suddenly there was a loud explosion, her big stack was forced back on the water and with a mighty roar the *President Lincoln* disappeared stem first under the waves. Eighteen minutes before we were a happy crew, proud of our ship, proud of the illustrious name she bore and of her record. Now all that was left of her was floating wreckage and in life boats and on rafts we were braving the Atlantic 800 miles from France.

After the ship sunk, the life boats started to take the men off the rafts into the boats. I was pulled into a life boat after being on a raft about an hour. The boat I was pulled into happened to be Captain Foote's. I didn't recognize him right away in his new guise, his blouse removed and in its stead a khaki shirt and a sailor's white hat pulled over his eyes.

We drifted around perhaps for about two hours when suddenly cheers were heard coming from the rafts quite a distance from the main group. Looking in their direction we observed what appeared on the horizon to be a French bark; it proved to be the submarine. From a distance, her wireless mast resembled rigging. When Commander Foote realized that it was the submarine, he gave orders to lay to, so we just drifted until she cruised into our midst displaying the German ensign. The submarine appeared to be about 450 feet long with a fully outfitted wireless mast. A couple of officers were stationed in the conning tower, and three or four of the crew attired in leather were walking about the deck.

The first raft she encountered was occupied by a seaman named Anderson, the ship's cobbler. Anderson was taken aboard the submarine and questioned by the German Commander as to the whereabouts of our Captain, and then taken below and given warm food and cognac. A baker named Chaddick, who also was on a raft alone near Anderson, saw him being taken aboard, and yelled to some lads near him, "Well, they don't get me," and swam from his raft to a more crowded section of the survivors. Cruising in among the life rafts and boats the German Commander kept inquiring for the Captain of the *President Lincoln.* Invariably every one replied that "he had gone down with the ship." Few knew that he had survived.

Apparently satisfied, the Commander then discovered Lieutenant Isaacs in a boat wearing his blouse, one of the very few officers who wore them that day. Ordering Lieutenant Isaacs' boat alongside the submarine, the German Commander pointed toward him and said: "You will come forward," Lieutenant Isaacs boarded the submarine and saluted; the salute being returned, his "gat" was taken from him and we realized that he was a prisoner. Anderson was then brought out and ordered into Lieutenant Isaacs' boat. Then an officer on the submarine produced a movie camera and numerous pictures, were taken of us.

The submarine then drew off a distance with her bow facing us, forward gun aimed at us and a seaman stationed apparently ready to fire on us. A deadly silence settled over us all. We felt that this was to be the end of it all; we were to be shelled. They had taken one of our officers prisoner and pictures of us. One "gob" remarked, "Well, here comes the fireworks." Just as we expected to feel the hail of the shells, the German Commander must have changed his mind and ordered the man away from the gun. She disappeared shortly afterwards, reappearing in about an hour, cruised about us once again and finally disappeared in the east.

After our Captain assured himself that she had gone, he gave orders to tie the twelve life boats by bow and stern line, and each boat was to take aboard as many men as it could possibly hold. Most of the life boats had 50 men in them besides towing life rafts. When we felt that the submarine had disappeared for the last time our spirits arose. We expected to be picked up by the destroyers the next day, anyway, so why worry. Night came on, the sky was spangled with stars, although it was quite cold. Everybody was trying to cheer everybody else up. At about 9:00 o 'clock it became real dark, and at ten-minute intervals Coston signals were lit in each life boat. The boys started to sing all the popular songs such as "Good-by, Broadway; Hello, France," "Over There" and "Keep the Home Fires Burning." We must have been a weird looking group away out there, but thus the hours passed.

Shortly after midnight a quick pale yellow gleam quivered a short distance away from us, and the next moment the destroyer *Warrington* was in our midst. It was the most welcome sight we ever saw, so again we cheered. Inquiring of Captain Foote when we had last sighted the submarine, she then told us to stand by to be rescued. This proved to be very risky work, taking so many men aboard in the darkness. Our boats would be brought alongside and we would stand up until the destroyer rolled towards us. Then the men on the destroyer would reach down and grasp our hands and pull us up. In that manner over 400 of us were taken aboard the *Warrington*. The destroyer *Smith* had in the meantime arrived and while the work of rescue was on circled about on the alert for any appearance of the submarine. She later took aboard the balance of the survivors, about 350.

When we got aboard the destroyer the crew showed us every courtesy. They fed us, brought forth dry clothing and shoes and gave up their bunks to us and went up on the decks and slept. Their cigarettes running low, each cigarette was cut in half and so we all managed to get a puff, at least. At about 4:00 A. M., all the life boats and rafts being empty, we started back to Brest.

At about 8:00 o'clock the next morning (Saturday) the destroyer *Smith* was seen to cut across our bow at a terrific speed, and then we learned that she had sighted a periscope. She fired and dropped some depth bombs and we wondered what would happen next. On Sunday morning we arrived at Brest and were taken aboard the transport *Great Northern* homeward bound.

On Monday morning the Court of Inquiry was held aboard and for the first time we learned who was missing among our shipmates. I think that out of 785, we lost 26 men, three officers. Lieutenant Commander Whiteside, Lieutenant Mowat and Ensign Johnson and 23 enlisted men. Not one of the Army passengers aboard was lost, though some of them were helpless and had to be assisted from the ship. As thrilling as the whole experience was, I would face another without fear at sea, because that day I saw how the officers and men of the Navy conducted themselves in the face of danger and it made me prouder than ever that I was an American.

THE LOSS OF THE *Covington*

The *Covington*, formerly the *Cincinnati*, a Hamburg-American liner of 26,000 tons displacement, was manned by a crew of 734 men and 46 officers, and had been provided to carry 3,500 troops. (Her capacity was later increased to approximately 5,000.) She made an excellent record as a troop transport and at the time of her sinking was on her sixth voyage, returning from Brest to New York.

The refitting and repair of this German ship for service as a naval transport reflected great credit on the Boston 'Navy Yard, as she sailed with her first load of troops just 90 days after the work had been begun. The engines had been badly damaged by the owners and it was due chiefly to the efforts of Commander Frank Lyon, U. S. N., the Yard Engineer Officer, that repairs to the machinery were so expeditiously effected. The electric welding of the damaged cylinders, which was an unqualified success, took 51 days.

The *Covington*, Captain R. D. Hasbrouck, U. S. Navy, in command, had sailed from France on June 30, 1918, in a convoy of eight transports, including the

Lenape, Rijndam, George Washington, De Kalb, Wilhelmina, Princess Matoika, Covington and *Dante Alighieri.*

Captain E. T. Pollock, U. S. N., commanding the *George Washington,* was the Group Commander, and on the evening of July 1st the convoy was proceeding in two lines under escort of seven destroyers, speed 15 knots, all ships zigzagging in two lines as shown by the accompanying sketch.

At about 9:15 P. M. lookouts on board the *Covington* sighted the wake of a torpedo heading for the ship, 200 yards on the port beam. The Executive Officer, Lieutenant Commander Marshall Collins, U. S. N., was on the port wing and gave the rudder order, "Hard right." About ten seconds later the torpedo hit at the forward engine room bulkhead on the port side, well below the water line, throwing a mass of water and debris high in the air. Shortly after the torpedo hit, gun No. 6, on the port quarter, opened fire on what appeared to be a periscope wake.

The mortally wounded ship took a quick list to port of about 20 degrees; then, as the water found its way across the ship, swung back to a five-degree list. The main engines were at once put out of commission. Lieutenant B. C. Edwards, U. S. N. R. F., Chief Engineer, soon reported to the Captain, who was standing on the starboard bridge wing at the time of the explosion, that the water in the engine room was at the tops of the main engine cylinders, and in the fire rooms at the top of the boilers, thus completely flooding the ship's two largest compartments.

The crew went to collision quarters to save the ship and also made ready to lower the boats in case it became necessary to abandon ship. There were no lights showing and the emergency lights on the battle circuits were kept out, in order not to show the position of the ship. Although the escorting destroyers had driven off the enemy by a depth bomb barrage, the ship with her motive power gone was in a precarious condition. She was gradually listing to port and giving evidence of loss of stability. There was also a possibility of the submarine firing another torpedo and quickly sinking the ship before she could be abandoned with a resulting heavy loss of life. The increasing list, moreover, would soon preclude lowering the boats. The Captain decided to insure the safety of the crew and to conduct such salvage operations as might be possible. Under trying conditions the entire crew were transferred quickly and skilfully to the destroyer *Smith.*

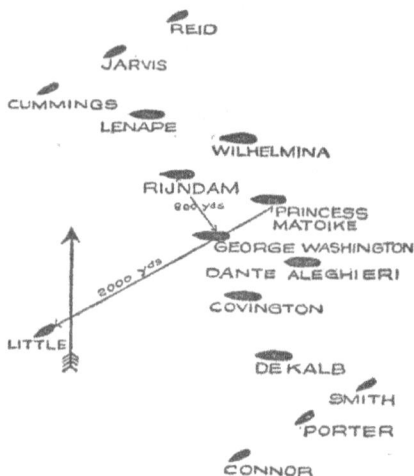

Torpedoing of the Covington

The Captain, with volunteer officers and men, remained on board to collect the ship's papers and to insure that none of the ship's confidential matter should be compromised. This work being completed, they then left the ship in the last boat.

After a careful search by the Smith of all boats and rafts, that had by this time drifted well to leeward, the *Captain* returned to the *Covington* with an organized salvage party in one of the ship's pulling boats and sent the heavily overloaded *Smith* to Brest.

About 5:00 A. M. the 2nd of July, the destroyer *Read* and three tugs arrived on the scene and a little later two more destroyers. About an hour later, the *Covington,* in tow of the tugs and protected by the destroyers, was headed at a speed of five and one-half knots for Brest, 150 miles away.

The ship was listing about 20 degrees to port and this list gradually increased. At noon the list suddenly increased about 10 degrees, and by 1:30 the ship was heeling 45 degrees to port and gradually sinking by the stern. Towing lines were then let go. At 2:30 her bow rose sharply in the air to a vertical position, and two minutes later she slid rapidly below the surface with a loud rushing sound of escaping air and a great upheaval of the water.

Of a total of 780 officers and crew, only six men were lost, less than one per cent. Of the passengers, all were saved. Mr. William H. Fulton, the only civilian on board the ship, addressed the following letter to the Captain:

AMERICAN EXPEDITIONARY FORCES
YOUNG MEN's CHRISTIAN ASSOCIATION
Headquarters 12 Rue D'Aguesseau, Paris.

July 6, 1918.

Captain R. Del. Hasbrouck,
   U. S. Navy.
My Dear Captain Hasbrouck:

As the only civilian on your ship, the *Covington,* when she was torpedoed, may I venture to express to you my profound admiration for the conduct and bearing in the hour of peril of officers and crew from yourself, our Captain, down to the humblest sailor?

In the enjoyment of the courtesy you so generously extended to me as Y. M. C. A. secretary, I had, about fifteen minutes before we were struck, completed the rounds of practically the entire ship, chatting with members of the crew as occasion permitted and observing with the interest of a landsman the appointments and routing of it all. It was shipshape, as you of the Navy would say; every man at his place of duty, the watch all alert, the gun crews prepared for any emergency. When the ship was struck, listed so heavily, shuddered, and to me seemed about to go down, we had the conditions which would have made for panic in any but the most thoroughly disciplined men. But I did not see even the suggestion of panic. Orders were handed down and obeyed almost as though it were one of our "abandon ship" drills.

Conduct in accord with the very finest traditions of the Navy was taken for granted. It was not exceptional, it was everywhere.

The most vivid pictures of heroism were rapidly impressed upon my memory — the gun crews firing away with unerring accuracy from gun decks that so far as the gunners knew might at any second be submerged; groups in the water, crowded on rafts and singing their merry songs, "Keep Your Head Down, Fritz," and "Hail! Hail! the Gang's All Here." Men voluntarily going back and forth between the destroyer and the ship, as many as three times, to bring their shipmates off, and only in utter exhaustion yielding the oars to other hands! It was nothing short of sublime, yet it all went on as if only a part of the regular routine. Danger was not only not feared, it was disdained.

Sad as it is that any of our brave boys should have lost their lives, it is remarkable that there were so few. It was one of those miracles which are performed only through the wisest forethought and the finest discipline.

It must afford you the deepest satisfaction. Sir, to reflect that in so signal a way you were not only the representative of our country but the arm of Him who travels all seas in bringing about His great purpose.

For myself I shall always be grateful that it was given to me in a humble sense to represent the nation's civilian population in an hour of strain and peril, and there to see the nation's seamen equal to that hour.

Deeply conscious of what we owe to you as an officer of our Navy, and no less appreciative of your uniform kindness to me personally, I am,

Respectfully yours,
(Signed), Wm. H. Fulton.

A youthful bluejacket of the *Rijndam,* Coxswain Baumann — one of the type who came in for the war and got what he came in for — told me the following story of the torpedoing of the President Lincoln and the Covington.

"It was the *Rijndam's* first trip across as a transport, and, believe me, all hands were on the job all the way to France keeping all their eyes on the lookout for submarines.

"But we didn't see any on the first half of the trip, and after the excitement of getting rid of the soldiers and taking a squint around Brest, we got under way for home expecting a quiet run without any rows.

"There were four of us left Brest at 10:00 o'clock on the morning of the 30th of May — the *Rijndam,* the *Susquehanna,* the *Antigone,* and the *President Lincoln.* The convoying destroyers left us that first night and the next morning we formed in line, the *Rijndam* on the left flank, and then the *President Lincoln* five hundred yards abeam of us, and then the *Susquehanna* and the *Antigone.*

"At about 8:30 that morning, having just come up on deck with my bucket of water, I heard the cry of 'Submarine!' come from the fo'c's'le. I looked forward, and there from right under our bows I saw come streaking the wake of a torpedo. And it was headed for the *President Lincoln.*

"I saw it hit her fair — right under the bridge. There was a boom, and then a great sheet of water and timbers and parts of the bridge flew up into the air.

"Then right up alongside us — too close for our guns — appeared the submarine, a long, green, slimy thing. She submerged immediately, before any one could have counted five.

"Right away the *Lincoln* started dropping back and listing to port.

"We were all watching her, of course, but we did not see the wake of the second torpedo. It hit the poor old *Lincoln* fairly well aft. There was another big boom and another shower of water and splinters and bits of boats — and over and above all flew the body of a sailor high in the air. It's going to be a long time before I can get that thing out of my mind — that tremendous shower of spray and wreckage and 'way above everything that poor smashed kid, his white suit standing out against the blue sky. I used to lay awake at night after I had turned in and wonder how he felt, if he felt anything....

"But to get back to the *Lincoln*. After she got the second torpedo, she straightened up a bit, her torn side gaping, her siren moaning, and her guns going full blast.

"We kept on. The *Susquehanna* and the *Antigone* beat it off to starboard at full speed. We felt like it was a dirty trick to leave the old *Lincoln* to wallow it out alone, but that's the way to play that game.

"From aft we watched her. She kept firing constantly — her guns pointing further and further downward as her bow lifted up. Her stem slowly settled. Boats and life rafts began to put off.

"In the distance we saw the last gun spit out what it thought of the Germans. Then the old *Lincoln* slipped beneath the waves.

"We went on. I looked around for my bucket, and it was gone. All of which goes to show you that no matter what happens in a large way in this world, you've got to keep your eye on your personal property or you lose out.

"On our next trip, we left Brest at noon on the day before the Fourth of July, with the *Covington* and a whale of a convoy. The *Covington* was second ship from us. The *George Washington* was 500 yards on our port beam, and the *Covington* was just beyond her.

"The destroyers were still with us that evening, and it had begun to get dark.

"Suddenly there came a green sky-rocket from one of the ships on our port quarter. Then a moment later there was an awful boom as the *Covington* got hers. She was hit in the port side in the bunkers, and clouds of coal dust and a great mass of coal flew up.

"We saw her take a big list to port right away. Her siren began moaning, and she started drifting back.

"Then the destroyers got busy, and depth bombs began to boom and shake us.

"The convoy broke up and beat it in all directions. For a little while the *George Washington* blocked off our view of the *Covington*. Then for a moment we had a final look at her. She had stopped — a poor, wounded, helpless, moaning thing, listing more to port every moment.

"We saw a faint light on her, which immediately went out.

"Then we slid off into the darkness and saw her no more.

"It was a funny thing about those two ships. Each had tied up to No. 2 buoy inside of the breakwater at Brest just before she started on her last trip. And each was on her sixth round trip as a transport. So, on our next trip, the old *Rijndam* had to draw No. 2 buoy. Naturally, we felt that it was all up, with wreaths on the grave and Uncle Joe and Aunt Mary coming 600 miles to Newark for the funeral, but we could swim; so we didn't care.

"But the *President Grant's* crew did not want to travel with us. They kept sending us signals that we were a jinx, but we didn't mind that.

"Well, they went right along next door to us, and nothing happened. I guess it was because we hadn't made six trips.

"Anyway, we didn't sink."

Chapter Nine - U-Boats Bring War to American Shores— San Diego Sunk by a Mine

In the Spring of 1918, as the weather improved, enemy submarines extended their activities further westward and the new type of large U-boat cruisers began to be heard from, finally carrying the war to our own shores. From this time on the danger zone extended all the way across the Atlantic. This necessitated some new dispositions, including the provision of destroyer and subchaser escort in the West Atlantic, and imposed longer strains and hardships on those engaged in the transport service, but the hope of the enemy that destroyers abroad would be recalled and the movement of troops and supplies delayed and thrown into confusion, was not realized.

When the much-heralded U-boat cruisers reached our coasts, we were ready for them. The result was that the damage they did was inconsequential, their activities were confined to the less frequented sea areas, while the main lanes of ocean traffic to our principal ports were kept free. Aside from laying a few mine fields along the coast between Eastport and Hatteras, and attacking several small vessels, their attempt on this side does not appear to have been serious. Not a schedule was broken, nor was the sailing of a troop transport delayed by their appearance.

The home guard anti-submarine fleet, including chasers, patrols, destroyers, submarines and aircraft, protected the Atlantic Seaboard. Even anti-aircraft guns had been mounted on the Palisades of the Hudson to protect New York and Bay points should any submarine have brought over seaplanes to be assembled on deck for the purpose of taking flight over New York to drop high explosive bombs. It is true that the enemy destroyed several fishing smacks, some schooners, a few barges and three small steamers, but these losses were insignificant as compared with the German threat that their U-boats were going to blockade our coast.

On May 25, 1918, the *U-151* suddenly came up out of the deep ten miles off Winter quarter light vessel and sank the American schooner *Hadington.* In June, 1918, the S. S. *Carolina,* owned by the New York and Porto Rico Line, was sunk off the *Carolina* coast and the passengers set adrift. This was followed by the sinking of some schooners, the small steamers, *Texel* and *Herbert L. Pratt,* and a few barges. Also a U-boat destroyed by shell fire the lightship at Fryingpan Shoals, Cape Hatteras.

On July 21, 1918, a German submarine attacked the tug *Perth Amboy* and four barges three miles off Orleans, Mass., at 10:30 A. M. The tug was burned and the three barges sunk. Several men were wounded but none killed.

The attack was witnessed by summer visitors to the Cape and by villagers, gathered by sounds of the bombardment. The following is quoted from the *New York Times:*

"No moving picture manager could have staged a sea battle more effectually for the summer visitors in this vicinity. Bathers were taking their moniing dip and scurried ashore when shells (from the submarine) splashed within a few hundred yards of them and many of the bathers watched the exhibition of German frightfulness from the beach. Automobilists stopped their machines on the brow of the sand hills and scores of cottagers did not have to leave their piazzas to see every detail of the fight."

Major Clifford L. Harris, commander of the Cape Cod Battalion of the State Guard, related to the correspondent of the *Times:*

"Two shots came upon the beach scattering the crowd. I do not think they were intentionally fired upon the beach but missed their mark or ricochetted from the barges. One shot struck on the shore at Nausett harbor, I am told. The whole affair lasted one hour."

It so happened that there was one schooner that never came in contact with or in sight of the U-boats. This was the mystery or "Q" ship, *George Whittimore,* a four-masted schooner whose innocent appearance cloaked the destruction she carried for U-boats. She had been fitted out by the Navy Department and sent out in disguise to seek the enemy, prepared to greet him with a rain of shell fire from her concealed guns. She worked in cooperation with American submarines and cruised up and down the Jersey Coast for over a month, but met with no success.

The enemy mining enterprise was more successful than either gun or torpedo attacks, inasmuch as it resulted in the loss of the *San Diego,* an armored cruiser of 16,000 tons, on the south coast of Long Island. The loss of life fortunately was small.

In connection with German submarine operations in the West Atlantic, it is interesting to recall the visit to this country of the ocean-going commercial submarine *Deutschland* and also that of the man-of-war submarine *U-53,* made before the War Declaration of the United States.

OPERATIONS OF U-53 IN WEST ATLANTIC, OCTOBER, 1916

On October 7, 1916, the German submarine TJ-53 quietly slipped into the then neutral harbor of Newport, E. I., exchanged official calls, bought the daily papers and departed that same day, strictly in. accord with the requirements of International Law.

The next day, October 8th, the U-boat sank five merchant steamers off Nantucket Shoal Light Vessel, namely the *West Point, Strathdene, Christian Knudson, Stephano* and *Blommersdyk.*

The *West Point* was sunk about 2:20 P. M., about forty-five miles off Nantucket Shoal Light Vessel. The weather was calm, sea smooth. The crew were given time to take to boats but were not able to save any of their effects. Thirty-three shots were fired into the *West Point* by the submarine and two time bombs were exploded alongside. According to the statement of the Captain of the *West Point,* these bombs were attached to the boat falls after every one had left the ship and exploded about a half minute after the small collapsible boat from the submarine had shoved off. As the Captain explained it, "They blew a hole in her side large enough to drive a cart through."

The *Christian Knudson* was sunk at 10:30 A. M. about thirty miles south southeast of Nantucket Lightship. One hundred and fifty shells and one torpedo were fired at the *Knudson* before she finally sank. The *Knudson* was built with many small tanks for gasoline and was loaded with gasoline. These sub-divisions will account for the difficulty experienced in sinking her.

The submarine commander gave a signed penciled statement to the Captains of both the *West Point* and *Knudson,* stating the time and position of the sinking of each ship and also the nature of the cargo. Both Captains stated that the submarine, with considerable difficulty, towed their boats with all hands in them to within easy visibility of the Nantucket Shoals Light Vessel.

The *Blommersdyk* was sunk at 8:12 P. M. about two and one-half miles east of Nantucket Lightship. One torpedo was fired by the submarine at 7:30 P. M. with small effect, the *Blommersdyk* listing only slightly to port. At 8:00 P. M. the second torpedo was fired and exploded with tremendous force, sending a column of water high above the vessel's masthead. Then the *Blommersdyk* began to settle by the stem and at 8:12 sank, stem first, with bow remaining out of water. The *Blommersdyk* was a vessel of about 9,000 tons, loaded with wheat and automobiles and bound from New York for Liverpool. The *Blommersdyk* was abandoned by her crew before 6:00 P. M.

The *Stephano* was sunk at 10:05 P. M. about six to eight miles northeast of Nantucket Lightship. Thirty shells were fired into the *Stephano* with apparently little effect and then the submarine fired a torpedo which struck about amidships. The ship broke in two and sank rapidly. All passengers and crew had left the *Stephano* before 7:00 P. M, and before any shots were fired into her. The *Stephano* was a passenger steamer plying between Halifax and New York. American destroyers were dispatched from Newport to rescue survivors.

There were no lives lost and no injuries sustained by any of the passengers or crew of any vessel sunk.

THE LOSS OF THE ARMORED CRUISER *San Diego*

The *San Diego* was the only large man-of-war lost by the United States Navy. She was an armored cruiser of 14,000 tons displacement, carrying a crew of 1,169 men, nine midshipmen and 49 officers. Her armament consisted of four 8-incli, twelve 6-inch and 22 three-inch guns.

On the morning of July 19, 1918, the *San Diego,* Captain H. H. Christy in command, was coasting down the Long Island shore making passage from Portsmouth, N. H., to New York City, from which latter port she was soon to escort a troop convoy. She was so near the approach to New York that the few men who had finished their watch on deck or below were cleaning up and shifting into *' liberty clothes," preparatory to leaving the ship upon arrival in order to get full benefit of the all too short stay in port.

The ship was steaming at 15 knots, zigzagging in a smooth sea under light southerly airs, visibility six to eight miles, when, at 11:05 A. M., she hit a submerged mine in waters to the northeastward of Fire Island Light Vessel and sank in twenty minutes.

The mine exploded well below the water line against the port side at the forward end of the engine room and felt like a dull heavy thud. It lifted the stern slightly and shook the ship fore and aft.

Captain Christy, standing on the top of the wheelhouse at the time, thought the ship had been torpedoed and immediately sounded to submarine defense quarters and directed the guns to open fire on anything resembling a periscope. Both engines were signaled full speed ahead and the helm put over in order to point the ship toward the nearest shoal water.

Unfortunately, however, the blow was suffered in the vessel's most vital part, and the Senior Engineer Officer, Lieutenant C. J. Collins, reported both engines out of commission and the machinery compartments rapidly flooding.

Headway fell off promptly and this precluded any maneuvering either to combat a submarine or to beach the ship, but the Captain still had hopes of remaining afloat.

The Executive Officer, Lieutenant Commander Gerard Bradford, made a tour of inspection through the lower decks, and reported to the Captain that although all hands had gone promptly to their stations and done everything possible to save the ship, still the water was fast getting the better of them.

As an instance of thoroughness in the performance of duty, Carpenter David Easdale, in charge of the after-repair party, was found by the Executive Officer, shortly before the ship turned over, on the berth deck composedly engaged in tightening the dogs on a water-tight door leading to a flooded compartment.

After a lapse of ten minutes from the time of the explosion the listing of the ship to port began to increase a little more rapidly. Captain Christy, seeing the ship was going to capsize, then gave the order for all hands, except the gun crews, to abandon ship. The gun crews were directed to stand by their guns until they could no longer fire and this order was carried out to the letter. Thirty or forty rounds were fired from the broadside battery at possible periscopes and wakes before the port guns were awash and the starboard guns pointed up into the air by the listing of the ship. The crews were then ordered to take to the water.

In the meanwhile the depth charges on the after quarter deck had been placed on safety. Later, when the quarter deck was partly submerged, Ensign J. P. Hillman, the Ordnance Gunner, showed presence of mind by going aft and doubly securing the forks in order to guard against explosions. Had any of these depth charges exploded as the ship sank, many casualties among the crew in the water would undoubtedly have resulted.

The evolution of abandoning ship was performed in a seamanlike manner. Upon attempting to use the boat cranes it was found that the electric current had failed, due to the flooding of the dynamo compartments. Because of the rapid listing of the ship and the loss of electrical power the larger boats could not be hoisted out. The life rafts, whale boats and dingies were launched by hand. These, with mess tables, benches, hammocks, and lumber, comprised the floating equipment upon which the crew abandoned ship.

The vessel was cleared as if at drill, the men going over the side by divisions. There was a moment of anxiety as the crew in the water scanned the sinking ship for their Captain. A shout from a raft, "There's the Skipper, I see his bald head," broke the strain and a cheer went up on all sides for Captain Christy. True to the tradition of the sea, he was the last man to leave the ship and stayed with her as long as possible.

With the eyes of the men in the water watching him, as the vessel slowly turned over to port, Christy first passed down from the bridge to the starboard superstructure deck, then slid down the ship's side to the armor belt, transferred from there to the bilge keel and finally, as the ship's bottom rolled to the surface, he dropped to the docking keel and from there jumped into the water.

The *San Diego* floated bottom up for a moment, then slowly sank, disappearing twenty minutes after the initial explosion.

From this time on the men took their mishap as an outing; shouts, cheers and laughter filled the air.

As no radio report had been sent, the Captain ordered Lieutenant C. J. Bright to proceed in a dingy to the Long Island shore to request assistance. Bright accomplished his mission and wireless messages for help were broadcasted from shore stations.

In the meanwhile, boats displayed the national colors, sails were hoisted to attract attention, and in a short time the steamships *Malden* (Captain Brown), *Bossum* (Captain Brewer) and *S. P. Jones* (Captain Dodge), ignoring the danger of a lurking U-boat, came to the rescue. By 3:00 P. M. all survivors were on board and the ships on their way to New York, which port they reached without further difficulty.

Incidents occur in such disasters that oftentimes do not reach the public. Pay Clerk Gagan, Acting Supply Officer, with his Chief Yeoman, George J. Meyers, took the water, each with a life preserver in one hand and money bags and valuable records in the other. They thus saved $20,350 of paper money and pay receipts amounting to $130,000, besides the payrolls and records.

In this connection it is interesting to note the American sailor's spirit of fair play with the government they serve. About $27,000 of pay receipts, representing the amount paid out in the current month of the disaster, were lost. When the survivors were gathered together they were asked to make a statement to the Captain as to the amount of pay covered by these lost receipts they had drawn and to sign duplicate receipts. Of the $27,000 all but $900 was accounted for by the voluntary statements of the men. Since there were six lives lost, it is a fair assumption that these accounted for the $900 and that all the survivors responded honestly and fairly without any man's taking advantage of the lost records to draw more money than was his due.

During the night of July 19, the various vessels which had picked up the survivors arrived at the Port of Embarkation, Hoboken, N. J., where the ever-ready women of the Red Cross met them at the piers and supplied them with comfort kits and hot coffee before they were taken aboard other transports at the docks.

The muster that night, verified by another the next day, showed a loss of six lives out of a crew of 1,184 officers and men. This remarkably small percentage of deaths testifies to the high state of discipline maintained on board.

Inasmuch as on the day subsequent to the disaster six German contact mines were located by our mine sweepers in the vicinity of the spot where the *San Diego* went down, it was concluded that a mine laid by an enemy U-boat caused the *San Diego's* loss.

Chapter Ten - Mount Vernon Torpedoed

The Mount Vernon was formerly the large German passenger steamer *Kronprinzessin Cecile,* gross tonnage 19,503. This ship will be recalled as the "Gold Ship," which, in the Summer of 1914, just before the outbreak of the war, sailed from the United States for Germany with a large consignment of gold. While at sea she received notification of Great Britain's war declaration and, being beset with British cruisers, she turned back, effecting her escape by taking advantage of a fog to slip into the small port of Bar Harbor, Maine, where she was interned. Later she was removed under United States Naval Guard to Boston, and upon our entry into the war was fitted out as an American transport.

On the morning of September 5, 1918, the *Mount Vernon,* Captain D. E. Dismukes, U. S. Navy, in convoy with the *Agamemnon,* accompanied by an escort of six (6) destroyers was about 250 miles from the coast of France proceeding homeward-bound from Brest at a speed of 18 knots. The weather was fine, the sea smooth and all ships were zigzagging. Suddenly a periscope popped up about 30 degrees on the starboard bow of the *Mount Vernon,* between the two transports, and about 600 yards distant.

Seaman E. B. Briggs, on watch at the *Mount Vernon's* starboard bow gun, immediately opened fire. At about the same time Chief Quartermaster A. W. G. Hines sighted from the bridge the wake of a torpedo coming straight at the ship. The Officer of the Deck, Lieutenant George W. Milliken, U. S. N. R. F., ordered hard right rudder, rang emergency speed, blew the whistle to indicate change of course and sounded the collision call. The vessel had just started to swing when the torpedo struck amidships, exploding with terrific force and throwing a huge column of water high into the air.

For an instant it seemed as though the ship had been lifted out of the water, men at the after guns and depth charge stations were thrown to the deck, and the shock was so great that one of the five-inch guns was thrown partly out of its mount. Men below, in the vicinity of the explosion, who were not killed outright, were knocked into temporary unconsciousness.

The torpedo hit fairly on a bulkhead separating two boiler rooms, and had blown open a hole 19 feet in diameter, large enough for a Fifth Avenue Bus to drive through. This resulted in rapidly flooding the middle portion of the ship from side to side, for a length of 150 feet. She almost instantly settled ten feet in draft due to the 7,000 tons of sea water taken in through the hole, but stopped there, indicating that the water-tight bulkheads were holding and leaving a margin of two or three feet before her buoyancy would be lost.

The immediate problem was to avoid a second torpedo. To do this two things were necessary; first, to keep the enemy below the surface and confuse him by attack with depth bombs and guns; second, to make more speed than he could make submerged and so prevent his trailing and attacking again after nightfall.

The depth charge crew consisting of Gunners Mates Lutomski, Nielsen and Duffy, who had been thrown down by the explosion, jumped to their feet, and under the direction of Lieutenant Myers, U. S. Navy, proceeded to drop a barrage of five charges, which exploded at regular intervals about 200 feet apart and 150 feet below the surface of the water. This was a neat piece of work, the evolution being performed exactly in accordance with existing orders.

The Gunnery Officer, Lieutenant Commander Doyle, U. S. Navy, had devoted much attention in preparing for just such an emergency as this, and it may well be that the depth bomb launching device, designed and installed by him, together with his well-drilled crew, saved the ship. At any rate, the effect was to make the submarine realize that the attack was being promptly and effectively met, and that his only chance of safety lay in immediate submergence.

THE BELOW DECK HEROES

The next step was to beat the U-boat in the matter of speed, and it would be impossible to give too much credit to the men below, who accomplished this by sticking to their posts in engine and fire rooms.

These men were put to a severe test. The terrific explosion was followed by instant darkness. There they were, with certain knowledge that they were far below the water level, enclosed practically in a trap, with only a long, narrow passage leading to the open air above, and the ship in imminent danger of sinking. The sound of hissing steam gave warning of the added threat of exploding boilers. It is to the everlasting honor of our Navy that not one man wavered in standing by his post of duty.

Due to the explosion, one-half of the boilers in the ship were instantly put out of commission, and the feed line in use as well as systems of communications to the engine room and lighting circuits were destroyed. Under the direction of the Chief Water Tenders, Firemen and Coal Passers coolly and promptly went about their urgent business. By means of holding burning coal in shovels up to the gauges it was discovered that the water in all the boilers had disappeared below the glass, thus indicating that the feed line had been cut. Quick action was necessary to avoid boiler explosion. All hands turned to and succeeded in quickly shutting off the damaged feed line, starting the emergency feed pumps in the fire rooms, and pumping salt water from the sea into the boilers.

The 150 foot amidship flooded section was between the engine room and the forward boilers, and the flanking athwartship water-tight bulkheads held. Fortunately, steam pipes leading from the undamaged boilers through this stretch of water to the engines remained intact.

Lieutenant Commander P. A. Guttormsen, U. S. N. R. F., Chief Engineer, took command in the engine room. Although the main engines were for a while slowed down to the extreme slow speed limit, they were never stopped; within twenty minutes steam pressure was being again built up, and within two hours the ship was making the remarkable speed of fifteen knots, which she maintained back to Brest.

In the meanwhile, the electrical gang under the direction of Lieutenant C. A. Kohls, U. S. N., was engaged in running electric feed lines down the fire room hatches, and in less than a half hour this auxiliary lighting system was in operation and an improvised telephone system had been rigged for communication between the engine room and forward fire rooms.

Commander Adolphus Staton, U. S. Navy, the Executive Officer, who had built up and perfected the organization, took charge of all dispositions below deck. The repair parties of carpenters and ship fitters under Lieutenant Almon, U. S. Navy, the Construction Officer of the ship, proceeded to reinforce with shores the athwartship bulkheads flanking the flooded compartments.

While this was going on, Chief Boatswain Louis Placet, U. S. Navy, and his gang were at work on the forecastle getting ready to place the collision mat.

All naval vessels are supplied with what is known as a collision mat and gear for handling it. This large heavily lined canvas mat is designed and rigged so that it can be hauled down the outside skin of the ship to any hole which may have been made below the water line by collision, shell fire, torpedo, or other cause, thus covering it as you would place a piece of sticking plaster over a cut.

In order to pull the collision mat down the side of the ship into position, it is necessary to pass what is called the dip rope over the bow, the bight under the bottom of the ship, leading the ends, one on either side, aft to abreast the location of the damage, so that by hauling on one side the mat attached to the other end of the line can be pulled down under the water. Two other lines, a forward guy attached to the forward corner of the mat, and the after guy to the after corner, are so led that the mat can be stretched tight and hauled forward or aft into position as may be necessary.

After the torpedoing of the *Mount Vernon,* in passing the dip rope aft, it fouled the starboard anchor. In order to clear it, Chief Boatswain Mate Lyons promptly went over the side on a bowline at considerable risk to himself. The presence of mind and cool daring shown by this man is typical of the American sailor, whose collective seamanship has been responsible for saving so many lives in this war.

Of course, in the case of such a large hole as the one made in the *Mount Vernon,* a collision mat would be of no use; but the size of this hole was not known at the time, and the Boatswain's gang went ahead to rig their collision mat exactly as if at drill. As has been explained, however, in this case the ship was able to stay afloat and proceed without stopping the hole and pumping out.

Under the direction of the Senior Medical Officer, Lieutenant Commander E. E. Curtis, M. C., U. S. Navy, the 153 wounded soldiers on board, most of them helpless cripples, were stowed in their assigned boats, with life belts on and bedding and blankets furnished, in readiness to abandon ship if this became necessary. The burned and injured men from the fire rooms were received in the sick bay and given care and attention. So great was the desire of these men to do their utmost that it was necessary for the doctors to hold some of them to keep them from returning to the fire rooms to assist their shipmates.

Thirty-five men were killed by the explosion, the bodies being recovered two days later after the ship had been put in drydock at Brest. One man died of burns a few hours after the explosion and another several days later, in the hospital at Brest, making a loss of thirty-seven, all of the Navy, out of a total of 1450 on board, including 350 army passengers, 100 of whom were sick or wounded. Eleven others who were seriously injured recovered.

The *Mount Vernon* reached Brest two hours and thirty minutes after midnight September 6th, where she was docked for temporary repairs. On October 28th she arrived in Boston for complete repairs, after which she was restored to service as a troop carrier, sailing on the 23rd of February.

The war nose of the torpedo which did the damage was afterward found in No. 7 fire room and a photograph of it together with one of the holes made by the explosion are shown in the accompanying illustrations.

LUCKY ESCAPES FROM THE FLOODED FIRE ROOMS

When the *Mount Vernon* was torpedoed, Charles L. O'Connor, Chief Water Tender, was in No. 8 fire room, one of the compartments flooded. The explosion of the torpedo threw him to the floor plates. He was choked by the gases from the torpedo and almost fatally burned by the flames driven from the furnaces, but had sufficient presence of mind to try to shut a water-tight door leading into a large 1,200-ton bunker. The door was damaged, however, and could not be closed.

O'Connor was then swept off his feet by the inrushing water, but luckily caught hold of the large ventilator leading from the fire room up to the open air. Into this ventilator the water lifted him and his calls for help were heard from above.

Connor describes his experience as follows:

I looked about and saw Kinch, Water Tender, standing by No. 17 boiler. I saw him start for the ladder. After that I did not know what became of him. *(Kinch was one of those killed.)*

I then tried to close bunker door No. 7, but she would not work. Then I heard the speaking tube ring. Just as it rang I saw right between the boilers what looked like a wall of water. It was about seven or eight feet high and came from both sides and gathered in front of No. 18 boiler. The water formed a whirlpool and boxes, shovels and everything were being thrown about. I bumped against two or three bodies in the grip of the whirlpool. I worked myself to the side of the ventilator and poked my head inside. The water was just entering the bottom of the ventilator. I groped around but could get nothing to hold on to. I kicked off my shoes and braced my back. I managed to get up about seven or eight feet into the ventilator, above the water. Then I commenced hollering.

It was a boilermaker that answered my call and a rope was lowered, but the rope was too short. Another rope was lowered but that too was too short. I was just about ready to take to the water. All my strength had gone. A rope struck me in the face. I seized it and tied it around my knee. They pulled me up and brought me to the sick bay.

H. S. Smith, Fireman 2nd Class, also had a narrow escape from the flooded fire rooms of the *Mount Vernon*. He had just dragged a bucket of coal from a starboard bunker and was standing in front of No. 18 boiler where he had dumped his coal on the floor plates in front of the furnaces. He was not standing far from O'Connor when the torpedo hit and the inrushing water swept him from the fire room through the very door O'Connor had tried to close, back again into the same starboard bunker from which he had just brought out the bucket of coal.

At first Smith thought he had been washed into the ocean, but as he collected his wits he realized he was in a bunker in which there were still bunker lights burning. He also found that he was not alone, there being three other men there with him. They talked over their predicament, noted the rising water, called for help without result and finally, with little hope of saving their lives, decided to shift for themselves.

Smith crawled through the athwartship bunker from the starboard to the port side. He was at the end of his rope and could go no further to escape the water. He had about given up, when his eye caught a bunker ventilator duct. The opening was rectangular and measured only eight inches by fifteen inches. Smith had no idea that he could get his 155 pounds through that hole but, as he afterwards put it, lie stuck his head up into this ventilator to get a few last breaths before being drowned. As the water rose he jammed his shoulders into the hole and to his surprise the pressure of the compressed air and water forced his body up, and so assisted he succeeded in raising himself about six feet in this ventilator when he was stopped by a bend in the pipe. He then called for help and finally was heard from the upper deck. Chief Engineer Guttormsen called down, "Who is that?" The reply was, "Smith." The Chief asked, "Are you hurt?" The answer came, "I don't think so." Then Guttormsen said, "Hold on and we will cut you out."

At first a hole was started in the barber shop, but it was soon found that this was not low enough and the rescue party proceeded to the troop galley on the deck below.

Chief Machinist Mate Hudson and Fireman 1st Class Follis, his rescuers, worked in steadily rising water, where they were being struck by heavy debris such as meat blocks. When they began cutting, the water was up to their knees. By means of hand chisels the German steel was cut through, but before the work was completed the water had risen shoulder high and the ship had begun to list to port. Every time she rolled Smith was covered with water. Finally, almost unconscious, his shoulders and hips badly skinned, he was pulled out. Had the rescue been delayed. Smith would have been drowned. The place where he was stuck in the ventilator was completely under water when the ship took her final position.

"Hard Luck Smith," as he was called, enlisted on board the *San Diego,* and on his first voyage that vessel was sunk by a mine off Fire Island Light Ship. His second voyage was the one of the ventilator episode on board the *Mount Vernon.* Smith modestly recounts his experience as follows:

I was standing in front of No. 18 boiler, close to the alleyway, when I heard a gun fired. Expecting more, I braced myself. Then the torpedo hit. Fire from the high doors of the boilers behind me burned my shirt. The next thing I knew I was being spun around and around in a whirlpool which the water had formed. I remember being carried up and then down again. The next thing I knew I was going down and under.

I was washed through a hole and it seemed that I came out in the broad daylight. I thought I had been washed into the ocean. I looked all around and decided to take my shoes off. I then realized that I was in a bunker. The lights were still burning in this bunker.

At that time the water was about a couple of feet between the upper floor plates and the deck. I reached for the plates on the ceiling and in a few minutes swung myself up. A little while later I met two other fellows. We sat there for a while and gathered strength. Another fellow came up named Crabtree. We all began to talk it over and see what to do. After a while we all stopped talking and did some rapid thinking.

I noticed the water was filling the bunker. I reached the ventilator and got stuck up to the waist. The water pushed me up into the ventilator more. The water was rushing up and down and went over me about six or eight times. I had to hold my breath each time. I started to holler, and I think it was about twenty minutes when somebody set to work to get me out. They started at the barber shop. Then they went to the troop's galley and started to cut me out. Later I was taken out of the ventilator and to the sick bay.

Patrick F. Fitzgerald, Fireman 1st Class, was another man who succeeded in getting out of the fire room opposite where the torpedo hit. The explosion threw him down on the floor plates with his feet extending into the ash pans of a boiler. The cold water revived him and in the dark he managed to get hold of a ladder. In passing along a grating he stumbled over the unconscious form of L. Vallin. Fitzgerald kicked Vallin several times, reviving him and finally succeeded in leading him to safety.

Chapter Eleven - The Work of The Cruisers

"Of sea-captains, young and old, and the Mates — and of all intrepid sailors; of the few, very choice, taciturn, whom Fate cannot surprise nor death dismay!"

Of these I write. These of the deep sea escort of the large convoys, who checkmated the German raiders and prevented surface attacks with guns by the big U-boat cruisers. Theirs was the constant and unceasing toil, in summer and winter. "Down the wet sea lanes, across the grey ridges all crisped and curled," as Kipling puts it.

Seven days of rest in port, then out again, mothering liners and pot-bellied merchant ships loaded with their invaluable cargo. The hard part of it was that they rarely sighted land on the other side but met the escorting destroyers far out from shore, when they had to turn around to buck the heavy

Nor'westers and so for home again, only to coal, have a little run on the Avenue perhaps, a look at the movies, then back again with another convoy.

Four of the armored cruisers were sent to Halifax to convoy the Canadian and English ships and these had the worst of it, for the Winter of 1917-18 was one of the severest ever experienced in the North Atlantic.

After driving through a gale in below zero weather these ships often became so covered with ice that they resembled icebergs hewed into the similitude of ships.

All of these cruisers, both the heavies and the lights, were very much in my thoughts during that memorable winter. On Christmas I sent out a radio to all the ships under my command, "A Merry Christmas to all, especially for those at sea." One of the Captains told me long afterwards that when he read that radio in the midst of a howling gale he said to the Executive Officer, "We are not forgotten; post this on all the bulletin boards. I want every man in the ship's company to read it."

Not much was heard of our cruisers during the war and yet neither the transports nor the destroyers were more actively engaged. Little attention was paid to those silently moving vessels covered with their confusing coats of camouflage paint, holding to the sea, weather-beaten decks stripped of all unnecessary gear, quietly coming and going, attending to their business of getting on with the war.

Generally speaking, the larger and faster cruisers of Squadron One were used to escort troop convoys and the smaller vessels of Squadron Two to escort cargo convoys. [1] All the deep sea escort duty for our troop transports was done by the cruisers of Squadron One, except that beginning September 9, 1918, battleships of the Atlantic Fleet were assigned for escort duty with the troopship and fast merchant convoys. This was to guard against expected raids by enemy battle cruisers.

When organized in 1917, Squadron Two of the Cruiser Force was placed under the able leadership of Rear Admiral Marbury Johnston. During the year July, 1917, to July, 1918, the cruisers of Squadron Two were engaged in escorting 54 convoys of cargo vessels, totaling approximately 1,073 freight ships.

From the beginning of the fiscal year 1918-1919, Squadron Two was engaged in escorting fast merchant convoys from the Port of New York, and continued in the work of escorting medium speed merchant convoys out of New York and Ne-port News at regular 8-day periods. With the advent of enemy submarines on our coast, the previous exacting duties of the cruisers were increased by the necessity of escorting transports from Newport News to the rendezvous at sea with the New York section of the group.

The cruiser was the shepherd, so to speak, of the convoy, and the enemy was always watching for one ship to straggle or stray the least bit from the flock, which meant the torpedo and the gun for the laggard. Almost invariably, disaster overcame the lone ship. A most striking case, in point is that of the *Ticonderoga.* There the cruiser *Galveston* had a large group of slow steamers to care for, and in looking out for the many her efforts to save the one were not successful.

In addition to escorting hundreds of thousands of troops and hundreds of cargo vessels, these cruisers maintained their regular schedules of target practice and drill, preparing for battle and training thousands of men of all ratings for transfer to newly commissioned ships.

Not only did the cruisers distinguish themselves by their navigation and seamanship under most difficult conditions, special mention should also be made of their noteworthy engineering performances.

The long voyages nearly across the ocean and return without a stop were accomplished only by rigid economy in the use of coal and water, and by keeping boilers, engines and machinery in tip-top condition. It was necessary to take on board as much coal as could possibly be taken on deck and in the fire rooms in addition to their bunker capacity quantities amounting to about 200 to 500 tons, varying for different ships, enough to take them 500 to 1,000 miles before starting on their bunker supply. Even then, during the turbulent winter months, there were many times of great anxiety before the ships got in. The Seattle had to put into Halifax on one occasion in a fierce gale with only 150 tons; but the St. Louis, perhaps, had the closest call of all when she arrived at Hampton Roads with only 10 tons in her bunkers.

Continuous cruising required that the work of overhaul and repair to machinery and boilers be done almost entirely by the ships' crews, Where all did so well, it is perhaps hardly fair to mention one, but the *Huntington* may be cited, as an example. From May 13, 1917, to December 29, 1918, she steamed 71,391 miles; the total number of hours under way was 6,455; 44,459 tons of coal were consumed; and during this entire period of war service no work was done to the machinery in the Engineer Department by outside service. That was a most creditable record in upkeep and performance.

Our cruisers were also assigned to patrol duty and sent on various special missions. They were always on the alert to answer a call for assistance whether ashore or afloat, and at the time of the Halifax disaster our ships were amongst the first to the rescue.

Rochester action against U-Boat

The cruisers were the ocean escort and as has been pointed out their chief duty was to shepherd their convoys, guard them against raider attack, and against gun attack by U-boat cruisers on the surface. Their deep draught and size made cruisers a good target for enemy torpedoes and it was a risky business for them to attempt the tactics used by light draught, quick-turning destroyers. The absence of the latter, however, sometimes required that the cruisers take the risk in order to guard slow moving freight ships. As an instance may be cited the useful service of the *Rochester* when her cargo convoy was attacked during the night of June 25, 1918.

In the evening of June 25, 1918, cargo convoy H. H. 58, consisting of 13 ships — speed 8 ½ knots, disposed in two lines of six each, with the 13th ship in rear of the center — was proceeding Eastward on a zigzag course in Lat. 55° 40' N., Long. 13° 05' W., about 250 miles from Marlin Head, under escort of the U. S. Cruiser *Rochester.* (Formerly the old *New York* of the so-called Flying Squadron of Spanish War days.)

The *Rochester,* Captain A. W. Hinds in command, was in station 1,000 yards ahead of the convoy. The sea was smooth, with a light breeze from SE by E. The moon was practically full. The combination of bright moonlight and twilight made excellent visibility conditions for submarine attack and the enemy was further favored by the slow speed of the convoy. The destroyer escort had not yet joined up.

At 9.42 P. M. the *Atlantian,* the "Van" ship on the right flank, suddenly opened fire, and at the same time a column of water shot up along her starboard side, followed by the dull report from the exploded torpedo.

Captain Hinds at once ordered full right rudder, sounded to torpedo defense quarters, rang up full speed and headed toward the submarine. In that latitude the four-hour night combined with the bright moon and long twilight made it imperative to attack and drive the submarine down or ship after ship might have been picked off at the enemy's convenience.

The Navigator, Lieutenant Commander Jules James, U. S. N., took his battle station in the forward Crow's Nest to con the ship for ramming and dropping depth bombs. As the *Rochester* turned, Liser, a seaman lookout in the Crow's Nest, sighted the submarine close to the *Atlantian* and 30° on the *Rochester's* starboard bow. He promptly pointed it out to James, who at once gave order through the voice tube to the helmsman, "Full right rudder, swing 30 degrees!" As the ship was swinging the submarine submerged and fired a torpedo at the *Rochester,* then distant about 500 yards. The wake of the torpedo was sighted on the bridge and Captain Hinds reversed the starboard engine to full speed astern, thus hastening the turning of the ship, with the result that the torpedo missed, passing about 30 yards ahead.

In the meanwhile the *Atlantian* was slowly sinking and her crew was abandoning ship, although her gun crews continued to fire occasional shots. At 10:00 she was struck by a second torpedo and sank five minutes later.

The *Rochester* then interposed between the submarine and the convoy and steered zig-zag courses at full speed. She also signaled to *Atlantian* boats by blinker light that arrangements would be made to rescue survivors. At 1:20 a British destroyer was sighted and having been signaled the position of the *Atlantian* boats, she proceeded to their assistance.

Lieutenant Commander C. S. Graves, U. S. N., Executive Officer of the *Rochester,* reported that at 11:12 P. M. a muffled explosion was heard, and at 11:17 a fire was observed on one of the ships of the convoy which burned brightly for about ten minutes, when it was extinguished. At this time it was dark and as the convoy was in some confusion Graves could not make out on which ship this fire occurred.

During the time that the submarine was delayed by the action of the *Rochester,* the convoy was able to get such a lead that it was not until daylight, when the destroyer escort was in sight, that the submarine succeeded in again overhauling the convoy. At morning twilight, a torpedo was fired presumably by the same submarine, at the *War Cypress,* the rear ship on the left flank, but it missed. The *Rochester,* which had resumed station at the head of the column, again turned toward the enemy, which was sighted porpoising, or awash, but before the *Rochester* could clear the convoy to begin firing, the submarine submerged. Zigzag course at full speed was taken up to cover the rear and left flank of the convoy.

A few minutes later, 1:45 A. M., a second English destroyer joined as part of the Eastern escort, which was reenforced to six destroyers by 3:00 A. M., when the *Rochester* parted company with the convoy and set course for Boston.

[1] See Cruiser and Transport Organization Sheet in appendix.

Chapter Twelve - Contacts of Transports and Cruisers with Enemy Submarines

Prior to May, 1918, except for a few isolated cases, enemy U-boats confined their operations to sea areas east of the Azores. The destroyer escorts, therefore, were in the habit of rendezvousing with troop laden convoys in the neighborhood of Longitude twenty degrees West; about here they also left the homeward bound transports to make the best of their way unescorted from thence Westward. During the first year of the war, cruising in waters to the Westward of the Azores was comparatively uneventful, although after the demonstration of the *U-53,* attacks had to be looked upon as possible at any time or place in the Atlantic.

In the Spring of 1918, however, contacts began to be made further West. Among the first of these was an engagement between the U. S. Troop Transport *Pocahontas* and a U-boat cruiser which in all probability was one of the first headed for our Eastern seaboard.

In the forenoon of May 2, 1918, the *Pocahontas,* Captain E. C. Kalbfus, U. S. N., commanding, was proceeding Westward unaccompanied. The weather was fair and the sea smooth except for white caps. At 10:20 A. M., when about 1,000 miles west of Brest, a large U-boat cruiser came to the surface astern, evidently having failed in an attempt to attack submerged with torpedoes. At a range of about 7,000 yards she lay to across the transport's wake and immediately opened fire with two high power guns.

The transport returned the fire with her battery, which unfortunately consisted of old type guns completely outranged by those of the enemy. Even at extreme elevation, the American gunners saw with chagrin and disgust that their shells fell short, so the Captain ordered full speed ahead and steered zigzag courses. The engine room responded in fine shape, and soon the ship's engines were making 80 revolutions, 16.7 knots, the highest speed she had ever attained under German management even when new. The surface speed of the enemy U-boat was estimated as 15 knots.

In the meanwhile enemy shots were falling all about the transport, exploding on contact with the water and showering fragments on board. Captain Kalbfus reported that for 15 minutes the ship was under a heavy fire, high explosive shell falling close aboard, ahead and on both sides. That no direct hits were made seemed incredible as the enemy deliberately lay to and fired both of his guns rapidly and accurately so that fragments were plentiful on the bridge and elsewhere.

At 10:40 fire slackened because of the increase in range, at 11:20 the last shot was fired, falling far short, and soon after that the submarine disappeared astern. All ships were warned by radio. It so happened that one of the recipients of this message was the U. S. Cruiser *Seattle,* then not far away and engaged in rescuing a disabled freighter, the *K. I. Luckenbach.*

THE RESCUE OF THE *K. I. Luckenbach*

The *Seattle* (Captain De Witt Blamer) had turned over her convoy of troop transports to the destroyers at the sea rendezvous in the Bay of Biscay and was headed Westward, when in the morning of April 27, 1918, she received the following SOS from the U. S. Army Freighter *K. I. Luckenbach:* "Engines completely disabled. Must be towed to some port for repairs. Give assistance as soon as possible. Position at 8 A. M. Latitude 46 degrees 26 minutes North and Longitude 23 degrees and 57 minutes West."

The *Luckenbach* was a little less than 200 miles away and the *Seattle* immediately went to her rescue. The approach was made with caution, the *Luckenbach's* identity being tested by various codes to guard against falling into a submarine trap. At about 4:00 P. M. she was sighted, but on account of a rough sea and southeasterly gale, she could not be taken in tow until the wind and sea moderated.

Early in the morning of April 28th, the tow line was passed and the *Seattle* proceeded at a speed of 4 knots to the Westward in order to get out of the submarine zone. At 9:30 that night the ten-inch manilla tow line parted. By daylight all was in readiness to try again and a fresh start made. Progress was slow, however, and six knots was the maximum speed that could be made. In case of U-boat attack the tow would have been at great disadvantage because of its slow speed and inability to maneuver. No destroyers were available for escort.

On April 30th, because of shortage of coal, the *Seattle* had only 1,200 tons on board, course was laid for the Azores. On May 2nd, the following radio message was received from the *Pocahontas,* a Naval Transport of the same convoy the *Seattle* had just escorted across: "Engaged enemy submarine on surface 11 A. M. 2 of May. Latitude 46 degrees 25 minutes North, Longitude 28 degrees 10 minutes West. Gun range ten thousand yards. No damage."

At noon, two days before, the *Seattle* with the *Luckenbach* in tow had passed only 42 miles to the Eastward of this position and must have missed contact with the U-boat by a narrow margin.

Good luck, however, continued to attend these two vessels and on the 3rd of May the *Luckenbach* having been transferred to two tugs from the Azores, was safely taken to port for repairs, while the *Seattle* proceeded on her way to the United States.

THE Henderson RAMS A SUBMARINE

On August 13, 1918, the *Henderson* was cruising off our Atlantic coast when, soon after midnight, the ship passed through a large oil slick which aroused some suspicion, but it was not until the next day that it was learned that this oil came from the tanker *Frank W. Kellogg,* which had been torpedoed two hours previously by an enemy U-boat. For some reason the *Kellogg* failed to send out an SOS.

Shortly after the *Henderson* had passed through this oil at about 1:40 A. M., August 14, 1918, Private Roy O. Hicks, Marine Corps, stationed as a lookout in the fore top, sighted a long dark object on the starboard bow about 500 yards distant and coming straight for the ship. This was at once reported to the bridge by Second Lieutenant E. O. Bergert, U. S. M. C, the officer in charge of the watch in the foretop.

At about the same time, Junior Officer-of-the-Deck Ensign R. McKay Rush also sighted the submarine, and Captain Sayles, who was on the bridge, ordered right rudder, sounded general quarters, and headed for the enemy. The submarine was maneuvering to fire a torpedo, but the ship's prompt maneuver frustrated the attack and the U-boat submerged. The swinging of the ship and the immediate diving of the submarine prevented the gun crews from getting in a shot.

It was not definitely known at the time whether or not the U-boat succeeded in getting under fast enough to avoid the *Henderson's* ram, but when the ship was next docked, it was found that her starboard bilge keel had been partly bent and broken. As there is no other explanation, it is believed that this damage was caused by striking the conning tower of the submarine as she was in the act of submerging.

The following excerpts from a subsequent report made by Captain Sayles after the Armistice is an interesting sequel to this attack:

CAPTAIN SAYLES' REPORT

From a statement made during a casual conversation recently held with some French officers in Brest, I learned that when the surrendered submarine *U-139* arrived in Brest there was a former member of her crew on board, a mechanic and an Alsatian by birth, who, at his own request, had been interned with his ship.

This Alsatian had told the French officers with whom I was talking that the *U-139* had encountered an American transport off our Atlantic coast, which had attempted to ram her, and had succeeded in breaking off both periscopes, so that for the remainder of the cruise the submarine was unable to attack while submerged.

This part of the story I verified by personally inspecting the *U-139*. Not only are the periscopes broken but the thin metal weather screen on the forward side of the conning tower was badly bent as the result of the collision.

The following facts are also known: That the *U-139* made but one cruise, which was to the Atlantic Coast in August and September, 1918; that after August 14, 1918, the *U-139* did not make any further underwater attacks, but was strangely occupied in attacking with guns and bombs barges, fishing and sailing vessels off Cape Cod; that on her return to Kiel she was laid up for repairs which had not been commenced up to the date of the Armistice; that the *U-139* was the largest of German submarines and was commanded on her only voyage by Lieutenant Amauld de la Perriere, one of the most successful and enterprising of U-boat Captains, who, upon his return, was given command of another boat in which he had just arrived off the Azores to commence a new cruise when hostilities ceased."

There seems to be evidence that the disabling of the *U-139* can be credited to the *Henderson* and perhaps this can be fully established when Perriere can be interrogated and access had to the German records.

CONTACTS MADE BY SHIPS OF THE CRUISER AND TRANSPORT FORCE WITH ENEMY SUBMARINES

Space does not permit a detailed narrative of all encounters between submarines and U. S. cruisers and transports. The official reports now before me supply convincing evidence of twenty-six contacts between enemy submarines and the ships in the force under my command during the war.

In addition to these there are reports of many more possible contacts in which the evidence was of sufficient importance to induce the commanding officers concerned to submit official reports. The characteristics of the U-boat, the stealthy approach, the underwater attack and escape by hiding made it difficult at the time to get complete data. Doubtless there were many attacks in which torpedoes missed, perhaps by a narrow margin, without periscope or torpedo being seen. On the other hand there were also countless false alarms wherein disturbances in the water made by fish, tide rips, floating spars, or other cause, were reported as possible periscopes or torpedo wakes. One Captain remarked that on the first voyage made by his ship, judging from the periscopes sighted by the lookouts, there must have been a picket fence of submarines stretched across the Atlantic.

It would be hard to say what percentage of these possible attacks were bona fide. Perhaps when access is had to the enemy files and all data is carefully collated showing where the various enemy submarines were, at the times of the reported attacks, a fairly accurate estimate can be made, but even then there will be no way of verifying the attacks made by the submarines which never returned to port.

Notwithstanding the incompleteness of the data now available, the following table is of interest as a conservative indication of the effort of Germany to stop the flow of our troops to France. This table includes only those reported contacts of which there is convincing evidence.

Submarine sighted on Leviathan's starboard quarter distant about 1,000 yards. Leviathan opened fire and Nicholson saw periscope and dropped depth bomb barrage.

Periscope sighted 100 yards on port quarter. Attack frustrated by gun fire and maneuver. Gunnery Officer reported that the second shot probably hit the periscope.

Date	Reporting Ships	Remarks
10:15 P.M. June 22, 1917	De Kalb Havana Seattle Wilkes Fanning	Attack defeated by gun fire and maneuver.
1:30 P.M. June 26, 1917	Cummings Birmingham Lenape Wadsworth	Cummings attacked with depth bombs. Débris indicated that submarine was damaged.
10:00 A.M. June 29, 1917	Kanawha Edward Luckenbach	Torpedo narrowly missed Luckenbach. Attack defeated by gun fire and maneuver.
6:48 A.M. Oct. 17, 1917	Antilles Henderson Willehad Corsair Alcedo	Antilles sunk.
9:25 A.M. Oct. 28, 1917	Finland Beauford City of Savannah Smith Lamson Preston Flusser Corsair Wakiba Alcedo	Finland sunk.

Date	Reporting Ships	Remarks
11:45 A.M. April 4, 1918	Mallory Mercury Tenadores	Attack defeated by gun fire and maneuver.
11:00 A.M. May 2, 1918	Pocahontas	Attacked by U-boat cruiser on the surface. No damage.
8:57 A.M. May 31, 1918	President Lincoln Susquehanna Antigone Rijndam	President Lincoln sunk.
7:16 P.M. GMT June 1, 1918	Leviathan Nicholson	Submarine sighted on Leviathan's starboard quarter distant about 1,000 yards. Leviathan opened fire and Nicholson saw periscope and dropped depth bomb barrage.
11:28 A.M. June 16, 1918	Princess Matoika	Periscope sighted 100 yards on port quarter. Attack frustrated by gun fire and maneuver. Gunnery Officer reported that the second shot probably hit the periscope.

Date	Reporting Ships	Remarks
11:40 A.M. June 18, 1918	Von Steuben	Enemy U-boat having sunk the Dwinsk an Army cargo vessel used latter's survivors in boats as a decoy. Fired torpedo at the approaching Von Steuben. Attack was defeated by gun fire and maneuver.
4:10 P.M. June 18, 1918	Tenadores	Periscope sighted 1,500 yards distant on port quarter. Attack frustrated by gun fire and maneuver.
9:42 P.M. June 25, 1918	Rochester Atlantian Convoy HH 58	9:42 P.M. Atlantian began firing to starboard and about the same time was struck by torpedo. Torpedo passed 30 yards ahead of Rochester's bow. Same avoided by maneuver. 10:05 P.M. Atlantian hit by second torpedo and sunk.

9:15 P.M. July 1, 1918	Covington Geo. Washington De Kalb Dante Alighieri Rijndam Princess Matoika Little Porter Smith Read	Covington sunk.
11:25 A.M. July 19, 1918	San Diego	San Diego sunk by a mine laid by enemy submarine.
1:13 P.M. July 20, 1918	Kroonland	Periscope sighted on port quarter about 800 yards distant bearing two points forward of port beam. Attack frustrated by gun fire and maneuver.
5:32 P.M. Aug. 13, 1918	Pastores	First indication of submarine was a splash about 2,500 yards on the starboard quarter followed by the report of a gun. With glasses a large enemy submarine could be seen lying awash athwart Pastores' course 6 or 7 miles distant, a little on the starboard quarter and engaged with two guns, apparently 6 inch. About fifteen shots were fired, none coming closer than 1,500 yards. Pastores fired nine rounds but was outranged, the shots falling 2,000 yards short. Submarine disappeared and Pastores continued on course.
1:40 A.M. Aug. 14, 1918	Henderson	Henderson attempted to ram submarine and probably damaged same.
8:30 A.M. Aug. 21, 1918	Orizaba Sibouey	Submarine sighted in act of submerging. Orizaba attempted to ram. Also dropped depth bomb.
6:52 P.M. Ship's Time 10:52 GMT Aug. 23, 1918	Pastores Wilhelmina Hull	Attack probably frustrated by zigzag. Wilhelmina tried to ram; Pastores fired one shot. Hull dropped depth bombs.
Between 22 and 23 hours GMT Aug. 26, 1918	North Carolina Brazilian S. S. S. Sobrol De Kalb Group No. 58	Gun fire of North Carolina and Sobrol and maneuver frustrated attack.

12 P.M. GMT Aug. 31, 1918 (about dusk in W. Atlantic)	U. S. S. Zeelandia	Attack frustrated by zigzag. Submarine broke surface showing conning tower at about 200 yds. just forward of port beam.
9:00 P.M. Sept. 1, 1918	Wilhelmina Pastores	Torpedo narrowly avoided by maneuver of Wilhelmina.
7:37 A.M. Sept. 5, 1918	Mount Vernon Agamemnon 6 Destroyers	Mount Vernon torpedoed but reached port.
4:25 P.M. GMT Sept. 16, 1918	U. S. S. Pastores	Submarine sighted on surface about 3 miles distant ahead. Submerged before guns could be fired. Attack evaded by maneuver.

Chapter Thirteen - *Orizaba* Depth Bomb Explosion— Great Northern Collides with British Freighter *Brinkburn*— Fire on Board the *Henderson*

The primary mission of transports was to land safely troops in France, but in so far as was consistent with the accomplishment of this mission the doctrine of the cruiser and transport force was to attack and destroy enemy submarines whenever circumstances permitted. Of course it was forbidden to run any unnecessary risk with troop laden ships nor was it expedient or wise to maneuver a large vessel up to a waiting submarine where the chances of the ship getting torpedoed were comparatively great and the chances of harming the submarine comparatively small; but, notwithstanding, there were occasions, especially after the submarines extended their operations all the way across the Atlantic to our shores, when transports and cruisers were able to use gun, depth bomb, and ram to good purpose, and if in most cases no material damage was inflicted, these attacking tactics at least had a wholesome effect on enemy morale.

A large transport is not as handy, however, in maneuvering into position to drop a depth bomb as is a quick turning destroyer, and to overcome this handicap Captain R. Drace White, commanding the *Orizaba*, and his executive officer. Lieutenant Commander W. P. Williamson, devised with most commendable zeal and resource a sort of howitzer for firing the bomb at the submarine.

A Lyle gun of the type used for throwing a shell with line attached was modified to throw a fifty pound depth charge. Both White and Williamson were Ordnance experts and they devoted much time and study to the development of this howitzer, which worked satisfactorily on their first test when a 50 pound depth bomb was successfully thrown about 150 feet.

On the next voyage, submarines were reported in their vicinity and preparation was made to use the gun in service, but it was first decided to fire one more test shot, this time with a somewhat more powerful propellant charge.

When all was in readiness for the test Captain White and Williamson stood at the piece as the latter fired the gun. Something went wrong and the bomb was prematurely detonated. The explosion at once killed Williamson and three men, S. T. Lambert, oiler, F. T. Mayer, baker 2nd class, and A. K, Baird, seaman, also wounding four other officers and twenty-two men, including Captain White, who fell to the deck with a broken jaw, broken knee and three flesh wounds. In addition to the loss of valuable lives, it seemed particularly hard that a few days after this regrettable accident the *Orizaba* reported a contact with an enemy submarine under such circumstances that the submarine might have been destroyed had the howitzer been in effective service.

Great Northern collides with British freighter *Brinkhurn*

Throughout the war the necessity of cruising without lights to prevent discovery by enemy vessels introduced a risk of collision which caused continual anxiety to those charged with the navigation of our transports, especially when loaded with troops. With these war requirements under certain weather conditions the best seamanship could not always avert collisions. They had to be looked upon as a necessary hazard of war.

There were many close shaves but, thanks to skillful ship handling, few disasters in the transport force, and even where collisions were inevitable the casualties and damages resulting were not great. In only one case was there loss of life amongst the soldier passengers; this was when the *Great Northern* collided with the British steamer *Brinkburn.*

At midnight October 2 to 3, 1918, the *Great Northern,* Captain S. H. R. Doyle, U. S. N., commanding, was eastbound loaded with troops, in company with her sister ship, the *Northern Pacific,* and the French chartered vessel *La France,* proceeding at 17 ½ knots, nearing the end of her voyage. Rendezvous with the destroyer escort was to be made that morning.

The last night on the edge of the war zone, before joining the Eastern escort, was always an anxious one. This particular night was unusually dark, sky overcast and air hazy, conditions handicapping U-boat attack but at the same time increasing navigational anxieties by making junction with the destroyers more difficult and also by increasing the likelihood of collision.

In peace time, under conditions of low visibility, ships in formation slow down, also running lights are always burning and searchlights frequently turned on, fog whistles are sounded and caution characterizes navigation. Things are different in war and the *Northern Pacific* was proceeding at high speed without a light showing, the eyes of her lookouts straining into the mist and darkness. Suddenly a dark form loomed up ahead and close aboard.

The Officer-of-the-Deck at once ordered the rudder hard over but collision was inevitable, and a moment later the two ships crashed together.

The ship on the westerly course was later identified as the British freighter *Brinkburn* and it was fortunate that she first struck against a heavy gun foundation on the starboard quarter of the *Great Northern*. The staunchness of this structure prevented serious damage to the deck and sides. She hit again, however, a few feet further aft and although with much less force, the second blow tore off the side plating for a distance of 25 feet, ripping a great hole in the *Great Northern's* side. The bow of the *Brinkburn* was crumpled up like paper, and as she backed off she left on board the *Great Northern* part of her stem, about 20 feet of her port bow plating, 10 feet of her starboard bow plating, one patent anchor, and 25 feet of chain cable.

Captain Doyle immediately took charge on the bridge of the Great Northern and all hands went to Collision Quarters. Many thought the ship had been torpedoed. The Army passengers were mustered at their Abandon Ship Stations and the crew proceeded about their duties of saving the ship in an orderly seamanlike manner.

The executive officer, Lieutenant Commander B. F. Tilley, U. S. N., took charge at the scene of the damage and under his direction the wreckage was cleared away, dead and wounded extricated and temporary repairs effected. This work was done in the dark. The ship's surgeon. Lieutenant Commander A. E. Lee, and his assistant, Lieutenant J. S. Callahan, with hospital corpsmen and stretcher men cared for the wounded. Repairs at the waterline were made by the Carpenter's gang under the direction of Construction Officer Lieutenant W. R. McFarlane and Chief Carpenter's Mate G. S. West. Oil was promptly pumped from starboard to port tanks and the ship listed to facilitate this work. So well was it accomplished that the *Great Northern* was able to proceed at 21 knots to her port of destination, which she reached without further incident.

It was unfortunate that the damaged compartments were occupied by troops, seven of whom lost their lives as a result of this collision. Five of these, Charles R. Mason, late private, U. S. Army; Osias Davidovici, late sergeant, U. S. Army; Darral Allman, late sergeant, U. S. Army; Harry E. Weigel, late sergeant, U. S. Army, and Marrion H. Worrel, late sergeant, U. S. Army, were killed instantly by the direct and immediate physical effect of the impact, which took place exactly where these men were sleeping; John E. Ransom, late sergeant, U. S. Army, died later of injuries so received; and Clayton W. Searcy, corporal, U. S. Army, was lost overboard, through the hole in the *Great Northern's* side, made by the collision. On the *Brinkburn* one man was killed and one man injured.

FIRE ON BOARD THE Henderson

Danger from fire and internal explosion was an ever-present menace. It speaks well for ship's guards and also the close surveillance maintained at the docks that there was, throughout the war, so little evidence of successful German plotting to destroy our ships. In the rush of transport loading it was no easy task to scrutinize every piece of cargo and lump of coal passed aboard to see that it contained no hidden explosive or infernal machine with time device to start a fire.

On board the *Pocahontas* six fires occurred within a period of three weeks while she was under repairs, but all were discovered and extinguished before serious damage was done. Other occasional fires were started in other transports but the efforts of the plotters in this direction were all abortive unless enemy agents had a hand in the *Henderson* fire, which was the only one of consequence.

The transport *Henderson*, Captain G. W. Steele, Jr., sailed from New York on June 30, 1918, and was proceeding in a convoy of 15 troopships escorted by the U. S. Cruiser *Frederick*, Captain W. C. Cole, U. S. N., Group Commander, and the destroyers *Calhoun*, *Mayrant*, and *Paul Jones*, when S. J. Cosick, EL. 1st class, on board the *Henderson*, reported a fire in a forward hold of that vessel. The cause of the fire was never discovered. No one had had access to the hold since the ship sailed.

Lieutenant Commander W. C. Barker, the executive officer, was first at the scene of the fire and under his direction the crew, armed with fire fighting apparatus, did everything they could to control and put out the flame. Smoke helmets, respirators, and wet towels, however, were of no avail against the heat and smoke, nor could the flames be checked by the numerous streams brought to bear. The fire spread to store rooms, the electrical shop, carpenter shop and crews' compartments.

As the fire approached the forward magazines the sprinkler safety device was turned on and the powder flooded to guard against explosion.

Air port gaskets were burned off. The ship settled about 8 feet by the head, due to the water that had been poured into her and this brought these damaged air ports under water with the result that additional large quantities entered the ship.

In the meanwhile the Group Commander, Captain Cole, had directed the transport *Von Steuben* and the destroyers *Mayrant* and *Paul Jones* to stand by the *Henderson*, and when it became evident that the latter would have to return to port, he directed the destroyers to transfer the 800 marines and 750 Navy passengers from the *Henderson* to the *Von Steuben*. By working all night this transfer was completed at 6:00 A. M. without mishap. The crowded *Von Steuben* then proceeded at 20 knots for France, while the Henderson, escorted by *Mayrant* and *Paul Jones*, headed for Philadelphia.

All that night and the following day the fire was fought. As the list of the ship exposed the damaged air ports, men were lowered over the side to batten them down in an attempt to limit the quantity of water entering the vessel. Pumps were kept going at full capacity and finally fuel oil was pumped overboard to relieve a situation which was becoming dangerous. By 9:00 P.

M. the list to starboard had increased to 14 degrees. As the ship was well down by the head, there was danger of capsizing in case the sea roughened, so Captain Steele ordered 150 men of the crew transported to the *Mayrant,* while those remaining on board continued to fight the fire and the excess water.

At about 4:40 A. M. the next morning, July 4th, the wind freshened on the starboard bow and although the Captain tried to head to the wind to prevent a sudden list to the other side, the ship suddenly rose on an even keel and then heeled to port. It was an anxious moment as no one could tell whether she was going to capsize or not. Luckily she fetched up at 22° to port and then steadied with a 20° list. The level of the water within the damaged part of the ship, at this time, was nearly even with that outside and the transfer of water from starboard to port, accompanying the shift of list, deadened all fire in the ship.

With the fire out, the work of controlling the water was simplified. During July 4th pumps, handy billies, and bucket lines were kept going at maximum capacity and the ship began to rise, gradually tending toward an even keel. By nightfall all danger to the ship had passed. The following day she arrived at the Navy Yard, Philadelphia, and went into drydock for repairs. These were speedily effected, and the *Henderson* was off again with another load of troops.

Chapter Fourteen - Sidelights on Transport Life, Embarking Troops

In the beginning it was attempted to conduct all troop movements in secrecy; the soldiers were taken from the camps to the docks and aboard ship as quickly and quietly as possible. Once on-board ship they were kept in their berthing compartments and not even allowed on deck until after the vessel had cleared port.

But as their number multiplied, speed in embarkation became a matter of great moment, secrecy was really out of the question; train load after train load of men from all parts of the country were deposited in Hoboken, the Northern Port, and in Newport News or Hampton Roads, the Southern Port, from which the transports sailed. No attempt was made at concealment and the transports passed out to sea with troops on deck and bands playing.

One rule, however, was preserved inviolate. At Hoboken General Shanks and myself agreed that on troop movement days no one should be allowed on the piers who was not in some way connected with the service. Against many protests this rule was maintained. Friends, relatives and sighteers, all had to be barred; necessity required that the troops should have the piers to themselves.

This rule did not apply, however, to the workers of the Red Cross Service. These devoted women were always on hand with hot coffee, sandwiches and cigarettes for the "boys," most of whom were leaving home and country for the first time — and some of whom were not to return. I often thought what the bright, cheerful faces of these devoted women must mean to those home-sick youngsters. Their work was beyond praise, for there never was a morn-ing so dark or a night so cold as to keep them from their self-appointed tasks. I believe the rule was that they were not allowed to talk with the men, but every man knows that a woman's smile at such a time is a benediction.

It must have been indeed with strange and varied feelings that these young men of America stepped over the gangway to embark on their great adventure. Thousands of them, of course, were from inland and had never even seen blue water before.

Proper handling of thousands of strange men on shipboard under the cir-cumstances of submarine warfare required system and organization to the last word. Quick tum-arounds were a feature of our Naval transports, and their record of men carried per month is indeed a marvelous one.

A ship being ready to receive troops, all gangways were rigged and at the end of each was stationed a Naval officer with cards and tickets on which, in concise form, were printed instructions for each trooper; where he slept, where he ate, his abandon ship station, and the rules he must observe. The soldiers then marched aboard in steady and continuous lines over all gang-ways. As they reached the deck they were met by sailors who conducted them to their berthing compartments and showed them their assigned bunks according to their respectively numbered tickets. In order to avoid conges-tion while embarking, soldiers immediately climbed into their bunks and remained there until the compartment was filled.

A compartment having been filled, each man in his bunk, the next step was to stow rifles and haversacks and to learn the prescribed routes with the rules of the road for using them to reach wash rooms, mess halls, and aban-don ship stations. In this way thousands of men could be embarked in the short space of one hour, and the soldiers at once plunged into ship routine, which had to begin immediately in all its varied phases.

In the meanwhile, the executive officer of the ship conferred with the commanding officer of troops and his assistants, explaining to them their administrative duties aboard ship. Details having been made. Naval officers instructed Army officers assigned respectively to guard, lookout, police, and commissary duty. Sentries were at once posted throughout the ship, patrols established, and the policing and messing details put to work.

With the submarines operating on this side it was necessary to have aban-don ship drill before clearing Ambrose Channel. For the average soldier the first forty-eight hours on board a transport constituted probably the busiest and most unusual period of his life.

Abandon ship drill, or "drowning drill," as the men called it, was a continuous performance until the soldiers learned to go quickly and quietly to their stations on deck, both by day or in pitchy darkness, at the sounding of the emergency signal.

It was the duty of the Naval men to rig out and lower all boats and rafts. Experience had taught that in saving life the main thing was to get all the floating equipment in the water and clear of the ship's side. The soldiers, each one wearing a life belt, could then go down the rope ladders, generously distributed along the ship's side, and man their assigned boats or life rafts. All transports carried an excess percentage of rafts so that in case half of the boats, due to list of the ship or other cause, could not be lowered, there would still be enough floating equipment for all hands. Sailors were also detailed to lower rafts over the side, and one or two were assigned to each raft, to take charge in the water and rig the tow line to be passed from the rafts to the boats.

An amusing incident about "drowning drill" was observed on board a certain ship. A colored trooper had been sitting on a big Carlin raft for several hours; when told to go below, he replied, "No, sah, my Capt'n give me dis heah ticket what reads foah dis raft, and I ain't goin' to leave it," and there he remained, possibly with the momentary expectation of being torpedoed.

The Naval crew was also detailed to take charge in any emergency, such as fire or collision, in which case soldiers were to stand fast and give the Naval men gangway to carry out their duties.

On approaching danger zones all hands were kept on deck as much as possible, day and night. Every man was impressed with the necessity of constituting himself an individual lookout and to report immediately the sighting of any suspicious object or condition in the seas. In submarine areas reveille was one hour before daybreak, in order to have all hands alert during the twilight period, which was most favorable for submarine attack.

Each of the army personnel was served out a life preserver when he embarked, and in the danger zone was required to wear it or keep it constantly at hand day and night. Those detailed to abandon ship on rafts carried two navy emergency rations and every man carried a full canteen.

The rafts were designed to keep the men clear of the water, but in actual practice they served as a buoy or marker to which people could cling while awaiting the rescue boats. All men were provided with life belts and the rafts were particularly useful in facilitating the work of rescue by gathering the men together in groups. To pick up a large number of scattered swimmers would be a lengthy and almost impossible task, especially by night. The rafts, in most cases, were shaped like elongated doughnuts, were in sets of three or four different sizes so that they nested one within another for stowage about decks. The outside circumference of the rafts was fitted with grab lines sufficient to permit from thirty to sixty men to hold on if closely gathered together. Inside the raft was stretched a net which prevented exhausted men from getting adrift even if they temporarily lost hold of the grab ropes. A sufficient number of boats for all hands could not possibly be carried by transports, but there were always enough boats to accommodate the feminine war workers, and the sick and wounded.

All life boats were provided with the following outfit securely lashed inside the boats: sails and spars; boat bucket for bailing; edible emergency rations; breakers of water; one heaving line with small life preserver on end; one set of oars and two spare oars, row locks with lanyards; one first aid package including tourniquet; one water-tight package of calcium phosphide; one boat hatchet; one oil tank and two oil bags; one compass; Coston signals; safety matches; oil lantern trimmed and filled.

In the danger zone life boats were kept lowered, if practicable, to the rail of an open deck at sufficient height to protect them from the sea.

For each boat there were detailed a naval officer, or a naval petty officer, and six of the naval crew, and this boat officer commanded in all matters pertaining to lowering, taking on board army passengers, handling and directing the movements of the boat after it was in the water.

The Navy personnel was impressed that the Army passengers were in their keeping and that their first duty was to provide for the safety of the men in khaki.

Of course, it was not pleasant for Army men, used to lots of room, to be crowded in what seemed to them stuffy holes and to be continually hounded by irksome submarine precautions, such as carrying life preservers and filled canteens, daily abandon ship drill, no lights, no matches, "You can't stay here, you must go there," but they knew it was only for the few days in transit and they took their temporary inconveniences in excellent heart, appreciating that it was all for their good and safety.

The soldiers had big things to look forward to on the other side and Navy men in the transport service regarded them with envy. Transport life was irksome in many ways, with nothing to look forward to except the possibility of receiving a "tin fish" in the ribs, perhaps a glimpse of a periscope, a few shots, some depth bombs dropped (result unknown); but never a chance for a real stand-up fight.

With the Transport Service it was the same old story. The anchor dropped in an eastern port; troops disembarked; cargo booms rigged; lighters came alongside; winches and whips began discharging cargo, — night and day the work continued. Then as soon as the ship was emptied and the wounded and other passengers received on board, it was up anchor, out lights, abandon ship drill, etc., etc., over and over again.

There was no growling, however, and although there was little chance for glory, there was some satisfaction in the knowledge that the Navy Transport Service was taking an indispensable part in rendering our army effective.

The problem of messing a large number of troops was given careful attention because it was essential that they be fed rapidly and also well. The decision was reached that the most practical method was somewhat similar to a continuous cafeteria style. Large ten-gallon aluminum containers were installed, and serving tables were especially constructed so that as the lines of troops marched in one door of the mess room and filed by the serving tables, each man's out-held field mess kit was filled.

The soldiers then passed on to high, narrow mess tables on which they could set their kit, while standing; up. As each man finished eating, he left the mess room by another entrance, near which washing-troughs with. hot water were provided for them to wash their kits. If a man felt that he needed or wanted a "second" he; fell in at the end of the line again and got it.

The galley arrangements as originally installed for passenger service were, of course, entirely inadequate for cooking enough food for the large number of troops carried, so batteries of steam jacketed kettles were installed together with large numbers of 80-gallon coffee urns in which large supplies of well-cooked food, coffee, tea or cocoa, could be prepared quickly for the hungry soldiers.

On most of the ships enough fresh meat was carried for all hands on the trip East, and for a few days in port; but the ship's company, officers and crew, became all too well acquainted with the good old standby, "canned corned beef," on the trip home.

To supply, on board the *George Washington,* for example, some seven thousand souls their daily bread necessitated the installation of bread-making machinery far in excess of the original plans of the ship, and the bakery forces worked in eight-hour shifts, twenty-four hours a day, turning out a daily output of from six to seven thousand full loaves of A No. 1 bread. It was good bread, too, so good, in fact, that passengers on the return trip from France, accustomed to French war bread, were under the impression that they were being served cake.

To look out for the personal wants of the sweet tooths of soldiers and sailors, four canteens were installed, and the quantities of candy, crackers, cigarettes, etc., bought were astounding. On this ship (*George Washington*) during one trip, the sales, at prices less than in any store ashore and as near cost as change could be made, were between $45,000 and $50,000, and upon arrival in France the unused candies, etc., were sold to the Y. M. C. A. or Naval Store at cost.

By regulations the Commissary is called upon to provide 180 different varieties of food. Handling 180 varieties of food in quantities that reach from 800 pounds to 79 tons speaks almost for itself.

It is a great sight on board a large transport to watch the almost unending lines of khaki file by for their meals. In spaces no larger than a private dining room at Sherry's they come by thousands upon thousands, and yet in such perfect order that in less than eighty minutes seven thousand soldiers have been served to the last man. The khaki line seems limitless, it may seem longer than that to those in the rear, but the coffee in the big pots remains hot, the stew continues to steam, and in less than seven seconds each man has his equipment piled with food. Perfect system and vast quantities of things to eat is the answer.

U. S. S. Leviathan

Special interest attaches to the *Leviathan* because she was the largest ship in the world, and for that reason was most useful to us as a transport.

Prior to the World War the *Leviathan,* then the German ship *Vaterland,* had been operating as a transatlantic liner between Hamburg, Cherbourg and New York. When war was declared she was interned at the Hamburg docks in Hoboken, and upon our entry into the war was seized and converted into a Naval transport.

The *Leviathan* was the only German vessel whose engines and machinery the enemy had not deliberately damaged, but even so it was found that after her three, years of idleness her great turbine engines were in bad condition, due to deterioration. In spite of the skill usually attributed to the German engineers, it was, moreover, found that these huge engines had also suffered from inefficiency in operation. In addition the piping, boilers, and auxiliary machinery of all kinds were in want of repairs.

Structurally the ship was in good condition and she required only the alterations to fit her for transporting troops, plus dry-docking and a thorough cleaning inside.

The excellence of the *Leviathan* as a transport lay in her great troop capacity and her high speed of 23 knots. Her great size and draft, however, were a disadvantage, as they prevented her from entering most of the British and Continental harbors and dry-docks.

On December 15, 1917, she sailed from New York to Liverpool with 7,250 troops on board. While at Liverpool the ship was dry-docked, and as a result of delay in docking and undocking the ship, on this first trip, remained in Liverpool fifty days. It was necessary to dock and Tindock on full moons as the highest tide was required to float the ship over the sill.

During this time the troop capacity was increased to 8,250 and upon her return to the United States this was further increased to 8,900 — on March 4, 1918, she sailed on her second voyage for Liverpool, but due to lack of water, poor berthing and coaling facilities in that port, she made her future voyages to Brest.

This huge ship has a cruising speed of 20 knots, burns 800 tons of coal a day and carries 8,800 tons of coal. In the early summer of 1918, with the urgency of hastening our troop movement overseas, her troop capacity was increased to 10,550. From December, 1917, to November, 1918, this one ship safely transported more than 150,000 troops to France.

There is a story about the Leviathan which is worth repeating here. On the 23rd of May, 1914, more than two months before war broke out, the *Vaterland* arrived in New York on her first voyage. She had been advertised extensively as the biggest ship in the world, and the finest. She was commanded by a. Commodore, and had four Captains of the German Naval Reserve as watch officers, etc. The occasion was celebrated by an official luncheon given on board and attended by the German Ambassador while the ship was tied up at the piers in Hoboken. At the time I was Commandant of the New York Navy Yard and was one of the luncheon guests. During the lunch I asked one of the officials of the Hamburg-American Line how many troops the *Vaterland* could carry. He replied, "Ten thousand, and we built her to bring them over here." He smiled when he said it. I replied, "When they come, we will be here to meet them," and I also smiled.

The next time I was on board the ship was three years later; she was at the same pier, she had a new name, she was flying the Stars and Stripes, and was being fitted out to carry 10,000 American troops to Europe to fight Germany.

SPANISH INFLUENZA EPIDEMIC

In fitting out transport medical departments, no expense was spared to make them as near to being real hospitals as possible. Each ship was fitted with a surgeons' examining room, dispensary, a laboratory, dental office, dressing room, operating room, special treatment room, sick bay and isolation ward. In addition to these, several dispensaries and dressing stations were established throughout the ship for minor cases, which the troop surgeons utilized for those patients not requiring sick bay treatment.

The Spanish Influenza Epidemic taxed the resources of the transport medical departments to the utmost. Although every effort was made to eliminate sick troops at the gangway, it was inevitable that large numbers of incipient cases were taken on board, and naturally the crowded berthing spaces favored contagion.

As an example, during the September, 1918, trip of the *George Washington,* although 450 cases and suspects were landed before sailing, on the second day out there were 550 new cases on the sick list. Entire troop spaces were converted into hospitals. Strict regulations in regard to spraying noses and throats twice daily and the continual wearing of gauze coverings over the mouth and nose, except when eating, were rigidly enforced. The soldiers were kept in the open air as much as possible, while boxing bouts, band concerts and other amusements on deck were conducted to keep up morale. The result was gratifying and the epidemic was soon under control. Admissions to the sick list were on a rapidly decreasing scale and although there were 131 cases of pneumonia and 77 deaths before arrival in Brest, still there were only 101 additional cases for the hospital and the remainder of the troops went ashore cheering and in fighting trim.

Computation of final tabulations from all ships show that 8.8 per cent of troops transported during the epidemic became ill, and of those who had either influenza or pneumonia, 5.9 per cent died. This gives an average Army death rate for the individual trips of 5.7 per cent per thousand. Navy morbidity rate was 8.9 per cent, and Navy death rate 1.7 per cent.

It is believed that these final statistics are highly favorable to sanitation on cruisers and transports, the morbidity and mortality being lower than in camps and civilian communities.

During this scourge in transports and cruisers there was a total of 789 deaths, and necessity required that many of the Khaki and the Blue be buried at sea. The following description of the ceremony of burial at sea was written by the Gunnery Officer of the *Seattle,* to send to the parents of a seaman buried from that ship early in the war.

WAR-TIME BURIAL AT SEA FROM THE CRUISER *Seattle*

The armored Cruiser *Seattle* was six days out on her third war cruise as ocean escort for troop convoy. News travels quickly in a ship, and before the morning muster at quarters we all had heard that one of the crew, ill of pneumonia, had passed away during the night.

The people of a ship are thrown intimately together on an ocean voyage and, in this case, war service added to the community spirit. The loss of our shipmate touched us all. Little was said but much thought was given as we assembled aft in answer to the tolling of the bell and the boatswain's pipe of the solemn call, "All hands bury the dead."

The service was conducted on the starboard side of the quarterdeck, the official place for ceremonies in a man-of-war. The bier was mounted outboard and draped with flags. Just inboard and forward stood the escort under arms. Space was left for the funeral party to march aft from inside the superstructure.

At the appointed hour, the ship's company, numbering about one thousand, ranged themselves in inverse order of rank around and abaft the turret guns. At the rail was rigged the gangway over which the body was to make its final passage from ship to sea.

The flag was then lowered to half-mast and the accompanying troopships in the convoy also lowered their ensigns to half-mast, thus joining in the ceremony, rendering homage in memorial of the life given just as truly in service for the cause as though it had been lost by the blow of a torpedo or an enemy bullet.

When all was ready the band played the funeral dirge, while the body bearers with the casket, followed by the pall bearers and Chaplain, marched aft at "slow time." The escort came to "present arms" and all hands stood at "attention" until the casket was placed on the bier and the dirge finished.

The Chaplain read the church services. At their completion the band played "Nearer, My God, to Thee." Then all hands "uncovered," the escort again came to "present arms," the Boatswain and his mates piped the side, and in reverent quiet — even the ship's engines were stopped — the body enfolded in the Stars and Stripes was committed to the deep.

Three volleys of musketry were fired, and the bugler ended the ceremony by sounding taps. The familiar and now mournful notes echoed in all hearts the call to the final sleep.

After a short pause the Captain gave the word "Carry on." The band struck up a march and the divisions went forward at "quick time" to their respective parts of the ship. Gun drills were resumed. Carpenters, shipfitters, black-smiths, and machinists picked up their tools. The propellers again churned the water, flags were masted, and the ship's work continued.

TRAINING OF HOSPITAL CORPSMEN IN NEW YORK CITY DURING THE GREAT WAR

The special intensive training of Hospital Corpsmen of the Transport Fleet was started in New York in the spring of 1917. This was largely made possi-ble by the ready cooperation and interest of Surgeon William Seaman Bain-bridge, U. S. N. R. F., of the *George Washington,* and Captain Pollock, from whose ship, the U. S. S. *George Washington,* the first corpsmen were sent for training.

The need for such a course was brought out clearly in May, 1917, when fifty-one new medical officers just entering the Navy were sent to New York from the Navy Medical School in Washington, for training. Therefore, in an-ticipation of the necessity of educating hospital corpsmen, a canvas was made at this time of the hospitals in and about New York City, to determine which ones could be best utilized for this purpose. There were natural ex-pressions of skepticism on the part of the medical authorities interviewed, as to the benefits to be derived by the corpsmen, because of their short stay on shore, but they were keenly anxious to be of help during our national crisis. In the beginning ten hospitals agreed to train the men, but before the cessa-tion of hostilities there were thirty City, State and Charity institutions (some took the initiative and requested of us the privilege of receiving some of the men) which opened their doors and gave instruction in the various branches of work as outlined by the Senior Medical Officers of the ships from which the men were sent. Practically all of these institutions provided lunches gra-tuitously for the men.

In all, the hospital corpsmen received training along the following lines: Dispensary work, including surgical emergency; pharmacy; general nursing work; elementary laboratory work, such as examination of sputum, blood count, etc.; Carrel-Dakin treatment and technique; operating room work; die-tetics; first aid dentistry; contagion; anaesthesia; X-ray work, and embalm-ing.

About 1,800 corpsmen profited by this course. It is an interesting note that some of them expressed the intention of using it as a basis for a medical career; others passed the New York State examination for licensed embalmers and are making this their life work. Many letters were received from the corpsmen in keen appreciation for all that was done for them and for the consideration they received on all sides.

Chapter Fifteen - The Loss of The U. S. S. Ticonderoga

The *Ticonderoga* was the former German steamer *Camilla Rickners,* interned at Manila, Philippine Islands, seized by the United States Government upon our entry into the war, and manned by the Navy Department.

She was a single screw steel vessel of about 5,500 tons displacement, speed 11 knots, and mounting two guns, a 3-inch forward and a 6-inch aft. The Naval crew numbered 16 officers and 108 enlisted men, while the Army passenger list on her last ill-fated voyage included 2 officers and 114 enlisted men. A grand total of 240 men on board, of which there were only 11 Naval and 14 Army survivors, the ship and cargo being a total loss.

The *Ticonderoga* sailed from New York on September 22, 1918, in a large cargo convoy of 24 ships under the escort of the United States Cruiser *Galveston.* The voyage was uneventful up to shortly after midnight of September 29th to 30th, when, although the speed of the convoy was only 9 ½ knots, the *Ticonderoga* began to drop astern, due to her inability to keep up steam because of an inferior quality of coal.

The convoy was formed in six columns, about 600 yards apart, and the distance between the ships in each column was about 400 yards. The *Galveston* was in station ahead of the middle column. No lights were being shown.

The night was dark, the sky cloudy, and it was beginning to get misty. A moderate sea was running under a gentle easterly breeze.

At about 2:30 A. M. the *Ticonderoga,* still dropping astern, lost sight of the convoy. The low speed continued for about two and one-half hours and for a short time was only 3 knots. By 4:00 A. M., however, she succeeded in picking up to 9 ½ knots, and effort was being made to rejoin the convoy, when, at 5:45 A. M., just before daybreak, visibility being about 200 yards, there was seen through the mist an enemy submarine bearing about 6 degrees on the port bow. She was lying nearly athwart the *Ticonderoga's* course, apparently with engines stopped.

Captain Madison, who was on the bridge, immediately rang up emergency speed, and altered course to the left to ram the U-boat. At the same time the general alarm for battle stations was sounded and the forward gun was ordered to fire at the submarine.

The enemy was on the alert, and within 30 seconds after being sighted fired a salvo of shrapnel from her two six-inch guns. At this point blank range the German aim was deadly accurate. One gun was leveled at the 3-inch crew on the forecastle and the other at the personnel on the bridge.

The submarine was so close that the 3-inch gun could not be pointed at it because of the interference of the deck, and before the American naval gunners could fire a shot, their crew was swept down by a hail of shrapnel. All of these brave fellows were killed at their stations, and their gun dismounted by the more powerful enemy 6-inch.

Upon sighting the *Ticonderoga* bearing down upon her, the submarine at once started ahead, threw her helm over, and quickly swung to the left, thereby escaping being rammed by the narrow margin of ten feet. She then turned again to the right, to cross ahead of the *Ticonderoga,* and although Captain Madison immediately shifted his helm, he did not have quite enough speed to reach the enemy, who managed to cross and again avoid the ram, this time by about twenty feet.

In the meanwhile, the U-boat was creating havoc by pouring salvo after salvo of 6-inch shrapnel into the American. The first shot at the bridge set fire to it, and killed all the men on it, except the Captain, Lieutenant Commander Madison, and Ensign Stafford, the Navigator. The former was temporarily stunned, however, by wounds in the face and shoulder.

The next salvo smashed the bridge and steering gear, and again wounded Captain Madison, this time knocking him off the bridge to the next deck and breaking his left knee joint.

It was probably the third salvo which dismounted the forward 3-inch gun and completed the destruction of the entire gun crew.

While this was going on, the *Ticonderoga's* after 6-inch gun had been manned and was ready to open fire, but deck obstructions prevented it being trained far enough forward to reach the U-boat. The latter was wary, and submerged while on the starboard bow before the after 6-inch could be brought to bear.

About ten minutes later the U-boat appeared again, this time two miles off the starboard quarter of the *Ticonderoga,* and resumed shelling the American, for the most part with shrapnel. The enemy's marksmanship was good, and the after gun crew was badly cut to pieces.

Although half of them were quickly killed or disabled, the survivors kept up a lively return fire, which finally drove the U-boat under the water again at about 6:15 A. M. The Americans were elated and thought a hit had been scored.

During the ensuing lull in the battle all hands turned to, putting out the fire and clearing away wreckage.

An early salvo had found the radio room, shattering the apparatus and killing the operator. This prevented sending out SOS signals.

The executive officer, Lieutenant Muller, had been awakened by a shell bursting in his room. As soon as he recovered his faculties, he found everything on fire about him, and only made his escape by dropping eighteen feet to the well deck.

He then went forward, and seeing the bridge demolished and the 3-inch gun dismounted with its crew strewn about it, he turned aft, and soon found the Captain, propped up on the after part of the midship upper deck, over the engine room, where, in spite of his wounds, he was conning the ship by means of a boat compass, and directing the work of putting out the fire and clearing away the wreckage.

Apparently the ship was not taking much water, and was still seaworthy and capable of being steered. Although the midship deck-house and bridge had been entirely burned away, the fire was now under control. Four boats had been burned, others riddled with shrapnel, and the entire upper works wrecked. A large number of men had been killed and wounded. Lieutenant Muller sent new details aft to replace the casualties in the 6-inch gun and ammunition crews, who stood by ready to open fire in case the submarine reappeared.

This happened about half an hour later, when the U-boat came to the surface off the starboard quarter, distance about 3 miles. Both sides again opened fire simultaneously. The submarine kept her decks awash, thus presenting a small target.

Not long after the renewal of the engagement, a 6-inch shrapnel exploded near the *Ticonderoga's* after gun, killing or disabling all except three of the crew. Lieutenant Ringelman then took station as gun pointer and gun captain and the survivors continued to serve the piece.

After the Americans had thus fired about ten shots, the enemy scored another costly hit; this time the shell exploded under the gun platform causing the training and elevating gear to jam.

After all efforts to make repairs failed, Lieutenant Ringelman, who seemed to bear a charmed life, reported to the Captain that his gun was out of action.

At this time, the U-boat, having reached a favorable position 1,000 yards off the *Ticonderoga's* beam, fired a torpedo which struck and exploded just abaft the engine room bulkhead. This was the *coup de grâce* and the ship began to settle rapidly.

After the 6-inch gun was silenced the enemy closed the range and kept up her deadly fire. Captain Madison, his ship a hopeless wreck, and his guns out of action, finally ordered the engines backed, to take way off the ship, and prepared to lower the boats.

Both quarter boats, filled with men, were wrecked by the guns of the submarine while in the process of launching.

After this disaster Lieutenant Ringelman started forward and found Captain Madison lying unconscious on the deck. He picked him up and threw him into a life boat amidships. This boat was lowered into the water without further mishap.

At about 7:45 A. M. the ship sank, stem first, from the effects of enemy gunfire and torpedo.

The submarine then cruised around, picking up vegetables from the wreckage, and finally went alongside the life boat, and demanded the Captain and Chief Gunner. The occupants gave incorrect answers, saying that these officers had been killed. They had previously removed their uniforms to avoid capture, Ensign Woodard and a soldier were summoned on board, and the German Captain, Franz, by name, ordered the life boat to tie up astern. As the U-boat started suddenly ahead, the life boat was only saved from capsizing by the parting of the tow line.

The submarine then went alongside a life raft which was crowded, for the most part with wounded men. Ensign Woodard and the soldier were put off on the raft, and Lieutenant Fulcher, the only officer in uniform, was taken on board. Lieutenant Muller had previously been picked up out of the water. With these two officers the U-boat steamed away, leaving the wounded men in the boat and on the raft to get along as best they might.

After several hours the life boat and the life raft were brought together and five men were transferred from the raft into the life boat by swimming.

All the men in the boat, except one, were wounded and unable to handle oars. Consequently, the boat and raft again drifted apart. Lieutenant Ringelman made sail and tried, until darkness overtook him, to get back to the raft, but without success. How many men were left on the raft is not known, but all were wounded. These poor fellows were never heard from.

After four days of indescribable hardship, the life boat was picked up by the British steamship *Moorish Prince,* and two days later, all the survivors able to stand the physical strain were transferred at sea to the British ship *Grampian.* Lieutenant Commander J. J. Madison, U. S. N. R. F., and four soldiers remained on board the *Moorish Prince.* All survivors were finally landed in New York City.

Out of the 240 persons on board the *Ticonderoga,* 11 Navy and 14 Army were saved. All of the Naval survivors, except one, had been wounded. I do not know the exact figures for the Army but most of them were probably wounded too.

Lieutenant Muller and Lieutenant Fulcher were taken prisoners on board the submarine, which was identified as the *UK-152,* commanded by Captain-Lieutenant Franz of the German Navy. Lieutenants Muller and Fulcher were repatriated via England by this same U-boat when she surrendered after the Armistice.

Chapter Sixteen - Foreign Transports in U. S. Convoys— Loss of *Dwinsk*— Adventures of Lieutenant Whitemarsh

Many foreign vessels were chartered by the United States Government to help carry our soldiers abroad and fifteen of these were assigned to my command, becoming practically a part of the Cruiser and Transport Force. They were issued the same sailing directions, were governed by our orders for Ships in Convoy, and operated at sea under the direction and supervision of the United States Naval Group Commander. These foreign transports were:

To make for smooth cooperation and to facilitate the communication of orders and instructions to the foreign Captains, each of these vessels carried a United States Naval Detachment consisting of one or two officers, a signalman watch for the bridge and a radio operator watch for the wireless room.

Ship	Nationality
Kursk	British
Czar	"
Czaritza	"
Dwinsk	"
Vauban	"
Caserta	Italian
Dante Alighieri	"
Duca D'Aosta	"
Duca Degli Abruzzi	"
Re D'Italia	"
America	"
Patria	French
France	"
Lutetia	"
Sobral	Brazilian

The Senior United States Naval Officer on board, was charged with seeing that proper zigzag courses were steered, the ship darkened at night, nothing thrown overboard that might point the trail, gun crews trained and alert, an adequacy of life saving equipment on board, and necessary emergency drills held against torpedo attacks, fire, and collision — in short, to see that proper measures were taken to safeguard the soldier passengers and to guarantee that the presence of the foreign ships would not prove a menace to the other ships in the convoy.

The officers called upon to perform this responsible war time duty, were young men of the rank of Ensign or Lieutenant, who, for the most part, had not been more than a year or two out of the Naval Academy. Less experienced in the ways of the sea and ships than the Captains with whom they served, they nevertheless understood the particular work in hand. Tactfully, yet firmly, they performed their duties in a thorough and officerlike manner.

The story of Lieutenant Whitemarsh, Senior Naval Officer on board the chartered British transport *Dwinsk,* may well serve to illustrate our type of Annapolis graduates.

The morning Whitemarsh reported to me for duty I was particularly impressed with his slender build and youthful appearance. I asked him how long he had been graduated from the Naval Academy. He replied, "One year, sir." "Do you think you could suppress a mutiny in a transport?" I inquired, and to this he responded, "Yes, sir; I've downed one and I suppose I could another."

At my request he then modestly recounted how he once boarded a schooner whose crew refused to get the vessel under way; how, with a few men, he had restored discipline, with the result that the Master had no further trouble in getting his orders carried out.

I listened with interest, and at the conclusion of his narrative gave him his orders to the *Dwinsk,* shook his hand, and wished him a pleasant voyage and a safe return,

THE Dwinsk TORPEDOED
A Cruise in an Open Boat

The *Dwinsk* sailed from New York on May 10, 1918, in company with thirteen other transports carrying troops, all of whom reached France in safety.

On the return voyage the ships separated before reaching this coast, and on the morning of Tuesday, June 18th, the *Dwinsk* was torpedoed by an enemy submarine in Lat. 38° 30' North, Long. 60° 58' West, some 600 miles distant from her destination, Hampton Roads, Va.

The torpedo was sighted at 9:20 A. M., 300 yards on the port quarter, "porpoising," that is, jumping out of the water at intervals as it raced for the ship. It was too close aboard to be dodged, and the torpedo struck abreast the after-hold. The Captain ordered the engines stopped, and the ship abandoned. No one was killed or injured by the explosion and no lives were lost in abandoning ship.

Nothing was seen of the submarine until, as the last boats were pulling away, it appeared on the surface some 3,000 yards away, and fired a number of shots at the ship. Her gunnery was very poor, even at this comparatively short range, and apparently little damage was done.

The boats clustered together a few hundred yards astern of the ship, and the submarine approached, keeping her two six-inch guns and four machine guns trained on them. The U-boat Captain then called all seven boats about him and questioned the occupants concerning the name of the ship, her destination, port of departure, tonnage, cargo and the nature of her duties. He made some effort to distinguish the Captain and officers, but they had concealed their identity by removing their hats and coats.

The submarine then, without taking prisoners, steamed off a thousand yards from the *Dwinsk* and again opened fire. At this range most of the shots were effective. One exploded the powder magazine, and the following one landed among the smoke boxes provided for making smoke screens. Great volumes of smoke arose shutting out the greater part of the sky. After the eighteenth shot, the ship listed heavily to port and at 11:15 sank, stem first, bow pointing skyward.

The seven boats made sail and headed to the westward. Lieutenant Whitemarsh, in boat No. 6, discovered that it was leaking badly and the sail, which was a lug rig, was found to be rotten and full of holes. There was no tinned meat in the boat, nothing but 24 gallons of stale water and some moldy sea biscuit. His 10 days' experiences are best told in his own words, which are quoted below from his official report.

STORY OF LIEUTENANT WHITEMARSH

"Our boat. No. 6, was sailing in the general direction of the rest of the boats, but losing distance steadily on account of having a rotten sail.

"Shortly after noon smoke was reported on the horizon to the Eastward. In a short time a ship appeared and developed into a four-stacker of the *Von Steuben* type. She was making full speed towards our boats and our wishes for an early rescue seemed about to be realized. But she suddenly stopped, avoiding a torpedo fired from the invisible submarine which was using our boats as a decoy. The ship opened fire on the submarine's periscope and fired five shots, the projectiles ricocheting over our heads. The ship then got under way quickly and soon disappeared.

The submarine came to the surface again over a mile astern, and approached our boat. She came alongside on our port hand and the Captain, who was burdened with iron crosses, asked us through his white-clad lieutenant what the name of the four-stacker was, and whether or not she was an auxiliary cruiser. I didn't know.

The presence of the submarine at such range gave an opportunity to study her characteristics. She was a dull slate gray in color, and showed marks of continuous running on the surface. The paint was worn off at the water line, where the hull was rusty. There was no lettering or distinctive markings on the submarine. She was about 275 feet long and had a beam of approximately 30 feet. Her armament consisted of two six-inch guns and four machine guns. The six-inch guns were situated midway between the conning tower and the forward and after ends respectively. The machine guns were grouped about the conning tower, two forward and two aft. The submarine was of the double hull type, with about five feet of free board. The tonnage was perhaps 2,500. The conning tower was directly amidships. If anything, the bow was a trifle higher than the stern. A life boat was carried, lashed to the deck, aft of the after gun. Still further aft there was an apparatus which I believe was used for mine sweeping or mine laying. Since it was housed it could not be made out accurately. At one time I counted thirty-seven men, including officers. The Lieutenant who acted as interpreter spoke broken English and understood with difficulty. The guns were kept trained on us while we were near the boat but they left us unmolested, not even inquiring as to our plans or provisions.

It was at this time that our boat started to pass boat No. 3 in a favorable breeze. Cadet Morrison shouted from boat No. 3 that we ought to stay together. Our sail soon developed greater rends which allowed Morrison's boat to forge ahead towards the leading boats, leaving us behind. It was a matter of indifference to us, except that a single sail might appear to a possible rescue ship more suspicious than a group of them.

We sailed all that night. The wind was ENE. Early next morning a heavy rain fell. The French sailor, Moellec, had oilskins, and three others had safety suits but the rest of the crew were thoroughly drenched. Two men particularly, who were in pajamas, were mercilessly exposed, even after those who were more plentifully supplied had shared their clothing.

We sighted a two-stacked steamer at dawn, close on our starboard hand. Showed a signal of distress, a red flare, but the steamer didn't reply to our signal. Five more times in the next four days we were passed by ships which we were almost certain would pick us up, but the period of jubilation invariably turned to one of despair when the ships headed away and left us. The *Von Steuben* had sent out a report saying that our boats were being used as a decoy by the German submarine, and this probably accounted for the failure of these ships to rescue us.

There was a heavy rain all day Wednesday, June 19th. At evening the rain lessened; the boat, now alone, keeping on the same course.

On Thursday nothing happened until evening when we sighted a steamer on our port hand, zigzagging. We showed several red flares but without result. At almost the same time we sighted a large bark, steering westward at such an unusual rate of speed that it was thought she might have been used as a supply ship for submarines. She showed no signs of having seen us.

On Friday we continued to sail on course WNW with a favorable breeze. Another steamer sighted failed to pick us up and we sailed through the night.

Watches were stood by every member in the boat. Three men were lookouts and stood two-hour watches. Currie (Cadet), Pritchard (First Officer), and I, took three-hour tricks at the helm in turn, while the remainder constituted the bailing detail, two men bailing for a period of a half hour.

As time went on signs of weakness began to appear; some were compelled to stop work, although they were still willing. The Maltese lad (assistant cook) named Sammut, had been torpedoed once before, when, in abandoning ship, he had been struck by the life boat swinging into the side of the ship. The injuries he had sustained to his hip had never completely healed.

Chief Baker Walker was given an additional allowance of water on account of the nature of his previous duties. The ration was a pilot biscuit a day and a half pint of water. Walker's mind, however, began to wander and he began to talk thickly of the coffee he was making and the pies he would be able to serve at five 'clock.

Spooner (fireman) went temporarily insane and in all my life I have never heard such an original and easy flow of profanity.

Early Saturday morning we sighted ship's boat No. 3 and went alongside. The crew had been picked up. Moellec (French seaman) entered the boat and did the greater amount of work in salvaging a new sail, a boat compass, a pair of shoes, can of biscuits and quantities of line, blocks and rigging. From this time the Frenchman was perfectly happy and busy, rigging an old shirt to a pole and running it up to the masthead for a distress signal, making capes from the old sail, making spray shields, splicing and working on the rigging. He never seemed to worry and was always ready with a smile and cheery word. His activity was unusual, considering that he was forty-five years of age. Since I was the only one who understood French, he used to talk to me for hours about his past life, and the weather.

By Saturday noon the wind from the east increased to a moderate gale. It was at this time that Pritchard, the First Officer, while having the sail reefed, allowed the boat to get into the trough. When I told him how to straighten out, he became angry and said he had forgotten more about sailing than I had ever known. A perfect accord could not be expected and certainly not enforced with the hatchet, our only weapon, so I allowed the matter to drop and took the helm myself.

All afternoon the wind continued to increase and the sea rose very high. The direction of the wind changed a bit to the right and held steady. The spray would occasionally drench us all. The sail, bit by bit, was taken in altogether. Two small triangles of canvas were rigged forward to keep her stem to the wind and weights shifted aft.

A line was made fast to the mast to indicate the direction of the wind, and I gave the helm to Seaman Fallon. He lay on his back in the stern sheets and steered while the boat was making five or six knots through the water. At 5:00 P.M. the gale was raging furiously with a heavy sea running. At 6:00 P.M. Fallon, drenched repeatedly, had a cramp and Cadet Currie took his place.

Currie was the 17-year-old son of a famous English sportsman and banker. He had not been at the helm five minutes before he saw a heavy cross sea coming down upon us. Unfortunately he released the tiller and obeyed the impulse to throw up his hands to keep the water off. The sea dropped in over the starboard quarter and washed him overboard, at the same time filling the boat to the gunwale.

I straightened the boat out, and all hands turned to with hats, buckets and shoes to clear the boat of water and to man the oars. The attempt to back the boat to pick up Currie only resulted in getting her into the trough. Currie was swimming towards us but not a third as fast as we were drifting. To save the lives of those remaining in the boat, we had to abandon the attempt to rescue Currie.

A little later another sea dropped down on top of the boat and knocked every one about, swamping the boat again. Pritchard, helmsman at this time, was suddenly stricken, and when the boat was again freed of water, he lay down in the bottom. I took the tiller and stood up in the boat in order to see the waves and feel the wind to better advantage. The men sat down in the bottom to improve the stability, and three of them appointed themselves my protectors by hanging onto my feet and knees. They evidently didn't want a second casualty.

The Frenchman stood up in the bow, like a gray ghost, hanging onto the mast. When the boat was poised on a wave, the bow down at an angle of 45 degrees and charging along at express speed, he seemed to be the least perturbed of the crew.

It was very dark and the wind, still increasing, brought intermittent rain squalls. This was not without advantage, since by opening the mouth water could be obtained. The water had a peculiar taste, as if there were quantities of ashes or dust in it. At times the rain would fall in torrents until the great waves were completely hidden by the rain splashes. This doubtless rendered the sea less perilous, a circumstance which perhaps saved the life boat from being wrecked.

It was about 11:00 o'clock that night when the wind began to shift rapidly. The wind would come from one direction and the seas from another. The waves were partially illuminated by a dim light, and this illumination was of great assistance in meeting them squarely. For fifteen minutes at a time I would keep the rudder hard right and then a few minutes hard left. In an hour there was almost a total calm, while the small boat tossed about aimlessly on the confused sea.

At first, when I made a remark about the wild beauty of the semi-illuminated sky and sea, the crew seemed to think that I had lost my mind. But after they heard about their unusual fortune in being at the center of a cyclonic storm and began to think about the tales they could tell when they landed, they began to cheer up and the conversation was quite lively. They forgot the incident of a half hour before, when one of the men, after a long and awe-inspired silence, moaned from the bottom of the boat, "Is there any hope, my good fellows?"

The calm was of short duration, however, and the wind set in again, bringing a torrential rain. The boat once more resumed its circling in the furious sea; the crew was drenched again and again with spray; the Frenchman stood at the mast and a detail of two men bailed out water without cessation.

After two hours of this, the wind steadied, though still blowing a gale. When it grew lighter in the morning, a long dark cloud was seen overhead extending across the sky from west to east, and when we were swept under it a chilly rain fell.

The wind coming from the west was dying down a little. My arms were aching after eleven hours at the helm, and after a sea anchor was rigged by lashing together two oars, the Frenchman relieved me. The wind moderated during the day, but the swell was high.

In speaking of the storm that day, Gregory, who had followed the sea for forty years, declared he had never seen anything like it. If, by having to endure the storm of that night again, the world would give him every luxury known to men for the rest of his life, he said he would refuse. He preferred the pleasures of a nice farm in Wales where he could spend the rest of his days with his wife and children.

Toward night we set sail heading southwest, the wind being northwest. At midnight the wind had dropped to a calm. Monday, Tuesday and Wednesday passed with light, variable winds and calms. These days taxed the courage of the men the greatest. They all knew we were in the Gulf Stream and drifting farther away from land every hour. When some of the crew, who had practically abandoned hope, began to sing familiar hymns, including "Nearer, My God, to Thee," I made them stop and the American seaman, Richards, and I sang "Homeward Bound," and other cheerful popular hits.

The food ration was cut to two-thirds of a biscuit a day with a quarter of a pint of water. The Second Engineer Officer, Pattison, became guardian of the hatchet, and whenever this weapon went forward to sharpen pegs or open tins, he would follow unostentatiously after and bring it aft again. He expected a raid on the food and water supply, but his fears were unfounded. The men were eager and prompt to execute every command or adopt every suggestion, particularly after the storm on Saturday night.

The spirit in the boat was excellent. Helpfulness and brotherly care were very evident in sharing clothing and sleeping places, and in assisting one another at work. Two of the weakest were excused from work. Those on lookout details had their eyes infected, until they were temporarily blind. Shirts were given as bandages and no efforts spared to make them comfortable.

Mother Carey's chickens, which followed the boat continuously, were looked upon as an omen of good luck. Small and varied colored sharks were called "land sharks" and an attempt made to spear them for food. Sea-gulls in flocks were considered a sign of proximity to land. Boxes, spars, and similar driftwood made the men happier. The first man to sight the steamer that would pick us up was to have the biggest dinner money could buy when we landed.

But the men were depressed in spite of it all. The sun would bake them mercilessly, and later, cold rains would chill them to the bone. One man made an attempt to drink salt water, and another thought it would be better to go over the side in the night and end it all. Discipline was insured only by the unchanging severity of command, combined with the proper regard for the welfare of the individuals in the boat. Moellec, Richards and Gregory were consistently cheerful.

Wednesday afternoon, towards four o'clock, the weather looked threatening and the wind increased. Rain began to fall very heavily. After washing the salt out of the sail, all hands drank their fill of water and caught an additional four gallons.

By midnight, the wind from ESE was blowing a gale with high seas and continuous rain. When we took a couple of seas the sail was shortened somewhat, but we made the most of the opportunity to run in. The crew was drenched with spray, but the time for compromise was past. Moellec and I relieved each other at the helm until Thursday morning, when the wind moderated and the rain stopped. It was calm all day.

A pleasing diversion during a watch was our time piece, a dollar watch marked "boyproof." It would run perhaps five or ten minutes at a time before it stopped. Shaking would start it again. The man at the helm stood very long watches unless he gave the "boyproof" his undivided attention.

Friday morning at 9:30, Collins jumped up and began waving his arms. He had sighted a steamer to the eastward heading towards us. The sail was left up until the hull and men of our boat could be clearly seen, and then we rowed alongside. It was the U. S. S. *Rondo*, Commander Grenning, U. S. N. R. F., in command.

Most of the men of the life boat were so weak that they had to be lifted up the sea ladder by means of a line, although a few of us managed it without assistance. The American sailor, Richards, who had sacrificed his rations to preserve his companions, was particularly weak. When I left the boat, two sailors from the *Rondo* were behind cutting holes in the hull and salvaging material such as oars, sails, water breakers and rigging. This was accomplished quickly and the boat left so that the next storm would knock her to pieces.

The survivors were given medical attention, clean clothing and food and shown every kindness human beings could bestow upon fellow creatures. The fearlessness of Captain Grenning in approaching the life boat when unarmed and when warned that the submarine was using our boats as a decoy, is most commendable and I am sure every survivor will remember him with infinite gratitude.

When picked up the life boat was 340 miles from Norfolk, Va. the *Rondo* reached port the next night, June 29, 1918. About six hours before landing, while standing near the bridge, I was presented with a paper which contained the following testimonial written and signed by all the survivors of the life boat.

"We the undersigned, survivors of the torpedoed steamship *Dwinsk*, wish to show our undying appreciation of the conduct of Lieutenant (j.g.) R. P. Whitemarsh, U. S. Navy, who, under the most trying and perilous conditions, set an example of courage and bravery beyond all praise, and we feel that his conduct and devotion to duty when face to face with destruction in a raging storm in an open boat, when most of us believed that the end had come, carried us through until the storm passed, and later, after many days in this boat, when all hope of rescue seemed small, he was always cheerful and hopeful, and encouraged us to further efforts."
(Signed)
    T. J. Richards, Seaman, U. S. N.
    R. J. Pritchard, First Officer.
    J. J. Skilling, Chief Steward.
    E. Griffith, Boilermaker.
    J. J. Martin, Barkeeper.
    C. Gregory, Linen Keeper.
    John Jones, Greaser.

John Wainwright, Donkeyman.
M. Keough, Fireman.
H. Spooner, Fireman.
"W. E. Soper, Storekeeper.
J. Sammut, Assistant Cook.
Je. Mouellec, Seaman.
James Pattison, Sec. Eng. Officer.
James Downie, Fourth Eng. Officer.
Dinsdale Walker, Chief Baker.
George Fallon, Seaman.
Harry Collins, Fireman.
James Wright, Barkeeper.

Von Steuben encounters submarine June 18, 1918

The *Von Steuben* while returning from France sighted a number of life boats on the port bow. Soon afterward a torpedo was fired, the wake of which was seen by an alert lookout when about 500 yards from the ship. His prompt report and the immediate maneuvering of the *Von Steuben* by the Captain saved the ship. Several depth bombs were dropped upon the estimated position of the submarine. As no SOS signal had been received at the time it was thought that the boats were nothing but decoys. Afterward it was discovered that they had been used as decoys but in addition contained survivors of the *Dwinsk,* torpedoed the day before. These were picked up by another ship.

Chapter XVII - Adventures of Lieutenant Isaacs, Taken Prisoner by A U-Boat

Lieutenant Isaacs was attached to the Naval transport President Lincoln at the time she was torpedoed early in the forenoon of May 31, 1918. Before the arrival of the destroyers which picked np the survivors during the night, while the *U-90* was steaming among the life boats and rafts searching for the transport Captain, the keen eye of the German Commander caught the stripes of Isaacs' uniform in the stem sheets of one of the life boats. The U-boat Captain put a megaphone to his mouth and sang out, "Come aboard!"

The boat ran alongside and Isaacs stepped to the submarine deck, and as he did so a German sailor relieved him of his revolver. (This was later returned to him.) Isaacs then made his way to the conning tower where he was given a glass of sherry and the Commanding Officer informed him that he was Captain Remy of the *U-90,* explaining in excellent English that his orders were to take the Senior Naval Officers prisoners whenever he sank a Naval ship.

After a half hour search for the *Lincoln's* Captain who escaped by disguising himself as a sailor, Isaacs said that he felt sure Captain Foote had gone down with the ship. The search was then abandoned and Remy ordered his prisoner below, where he was given warm clothing and allowed to lie down in one of the bunks. The U-boat then turned to the northeastward and proceeded at five knots to her cruising ground, which was about 300 miles west of Brest, arriving there on the following day, June 1st.

The following is a *precis* made up of excerpts from the official report of Lieutenant Isaacs:

"Early in the morning a radio was intercepted stating that the survivors of the *President Lincoln* had been picked up and that only a few were missing. That afternoon we sighted two American destroyers. They were so far away that Captain Remy thought that by heading away he could avoid being seen. He did not reckon, however, on the keen eyesight of the American lookouts. The destroyers instantly sighted him and gave chase.

We quickly submerged and a few minutes afterwards we felt depth bombs exploding all about us. Twenty-two bombs were counted in four minutes; five of them were very close, or seemed so to me, for they shook the vessel from stem to stern. To escape them we were making our best speed, zigzagging, and apparently doubling back on our course. The Petty Officer at the microphones, listening to the propellers of the destroyers, reported continuously whether they were getting closer or farther away to the Captain, who was in the conning tower. Soon they could no longer be heard, but we remained submerged at a depth of sixty meters for about one hour longer. Then Captain Remy brought his boat to the surface and continued cruising up and down at five knots speed.

The following morning, June 2nd, another American destroyer was sighted, but so far away that we were not seen. Remy then told me he felt that things were getting too warm for him in that vicinity and he intended to return to his base. We headed northwest and continued along the west coast of Ireland all that day and the next.

On June 4th, early in the morning, they called me to go hunting. We had approached a small island called North Rona, west of the Orkneys, where Remy was in the habit of stopping on each trip, weather permitting, to shoot wild sheep which were the sole inhabitants of the island. It seems that years before a hermit had come to live there and had begun raising sheep, which, after he died, continued to thrive. I counted 150 of them from the deck of the U-boat, for, after getting me up, the Captain changed his mind and decided that I was not to go hunting after all.

He sent one of his officers and two men in the small bateau which was carried between the inner and outer hull of the submarine, to the beach, and a few minutes later we could see them mounting the side of the cliff. I watched from the deck of the submarine through my binoculars. They shot nine sheep, one of which fell over the top of the cliff and into the water. Telling me that he knew he was a fool to do such a thing, Remy backed his submarine to within three feet of the cliff in order to pick up this sheep. One of the sailors pulled it aboard with a grapnel. A few hours later the hunters with the other sheep they had killed returned on board and we proceeded in a northeasterly direction around the Shetland Islands.

On the 6th of June we passed along the coast of Norway. The next day we got in touch with another U-boat which was running short of fuel. Her Captain was on board that night and talked a. while with Remy before returning to his boat lying a few hundred yards away. It was rather rough, so he did not take fuel from us but said he would try to make Kiel with what he had.

The following day, June 8th, we passed to the northward of Jutland into Skaggerrack, hugging the Danish coast. That morning we fell in with another U-boat, and for three hours both submarines maneuvered at high speed over a measured course between a lighthouse and a fixed buoy. (In submarine navigation, especially when maneuvering into position to attack, accurate data as to what speed is being made according to engine revolutions, is important, and these submarines were evidently engaged in checking their standardization curves.)

About noon time we entered the Kattegat. I had asked Remy if he ever rested on the bottom. That afternoon he submerged and rested on the bottom for about three hours. He told me that the submarine which was short of fuel had asked for assistance and Remy went to her aid, giving the other boat the fuel she needed during the night.

On June 9th we continued on our way and about 11:00 P. M. I was allowed on deck to smoke. I found we were in a little bay apparently with the lights of Sweden on one side and those of Denmark on the other. Although the sun had long since set, it was still twilight. (At that time of the year there is practically no night in this latitude — at least no real darkness.) We were at a submarine rendezvous, because I saw a second submarine about a quarter of a mile away and another soon came to the surface, making three in all. Finding that I was not far from a neutral country, I determined to try to make a getaway.

I had my life jacket which had never been taken from me and was hoping that it would get dark enough so that I could not be seen in the water. While I was moving over to the platform abaft the conning tower a German destroyer was sighted bearing down on us from the east at high speed. She was making the rendezvous in order to escort us through the Sound. Just as I was planning to slip over the side, Remy, who was never more than two yards from me, ordered me below. Before I passed through the hatch, I took one last look around and saw that the destroyer was placing herself at the head of the column and we were proceeding westward. Early the next morning I was on deck and found that we had passed into the Baltic and were heading in a southwesterly direction.

Before reaching Fehmarn we passed the battle cruiser *Hindenburg* and two other battle cruisers of the same type, also four armored cruisers, holding individual maneuvers.

We entered Kiel harbor, which was protected by a net, at 3:00 P. M., June 10th, and tied up at a landing near the entrance to the canal. Here I was allowed to go ashore for a few minutes' walk with one of the officers and I noticed probably a dozen destroyers in the harbor and about eight submarines of the same type as the *U-90*. In addition to these there were two large submarines probably 350 feet long, each painted a dark green and mounting a six-inch gun forward. These, Remy told me, were the new mine layers. At seven o 'clock we shoved off and in company with another submarine proceeded down the canal.

When I came on deck the morning of the 11th, we were in the Heligoland Bight. A Zeppelin was patrolling over head; and about nine o 'clock we passed a division of battleships, two of them being the Grosser *Kurfürst* and *König II.* They were sailing north at high speed, escorted by four large destroyers.

After passing through the locks at Wilhelmshaven we tied up alongside the mother ship *Preussen* and I was sent on board of her and put in a room with a barred port, the door locked and an armed sentry placed outside. We were lying in some back water from which it would be impossible for me to escape to the mainland; even had I done so I would have had to pass through the "most intensely guarded city of Germany" as they call it. One of the German officers told me it was practically impossible even for him in uniform to get out of Wilhelmshaven without passing through an enormous amount of red tape.

The *U-90* is a submarine built in 1916, approximately 200 feet long, carrying two 10.5 c. m. guns — one forward and one aft of the conning tower. Captain Eemy boasted that he could make 16 knots speed on the surface, and that he had demonstrated the superiority in speed that German submarines have over the American submarines when, some time previously, he had had an encounter with the *L-4;* that they had maneuvered in trying to get a shot at each other; that both submerged two or three times; and that finally he was able to fire a torpedo at the American submarine after getting into position, owing to his superior surface speed; that just as he was firing, the *L-4* dove and his torpedo passed a few feet over her.

While I was aboard we never submerged to a depth greater than 70 meters, although Captain Remy told me he could go to 100 meters. That last day, while passing through the Kattegat, when we were submerged for over 10 hours, we traveled most of the time at a depth of 70 meters. He seldom made more than eight knots speed submerged — I doubt if he could make much more. He carried a crew of 42 men and four officers. Another officer, Kapitan-Leutnant Kahn, was aboard for purposes of instruction, having had his request granted to command a submarine of his own. While I was at Wilhelmshaven, Kapitan-Leutnant Kahn came to see me in prison and told me he had just received orders to proceed to Kiel and take command of one of the new submarines.

Of the crew of 42 men, two were warrant officers — one the navigator, the other the machinist. The Captain's three assistants were lieutenants corresponding to our grade of ensign. One was a Naval Academy man who entered the Navy in 1913 — he was a deck officer; another was a reserve ensign from the merchant fleet by the name of Wiedermann, who spoke English very well, having been in America and England in peace times on various steamers; the other officer was a regular who had gone to their school for engineers and who was responsible for the efficiency of the machinery; he did not stand deck watch. The watch on deck was stood by the navigator (Warrant Officer) and the two ensigns (Leutnants). The Captain, Kapitan-Leutnant Remy, took the conn when ships were sighted and in passing through narrow waters. He had entered the Navy in 1905 and had traveled considerably, having been to America in 1911 on a cruiser which put in at Charleston, South Carolina, and into New York, in both of which places he had been hospitably entertained. He liked America but could not understand why America had entered the war. He believed, as all Germans are taught to believe by the governmental propaganda, that our entry into the war must have as its motive the rendering safe of the millions we loaned to France and England earlier in the war.

When I was captured the Germans were nearing Paris. On the submarine we received radio reports every day and it did look bad for the Allies. Remy and his officers were absolutely confident that the war would be over in a few months, and would end in a big German victory, for as they said:

"France will soon be overrun by our armies and there will be no place for the American troops to land. Besides, you are coming over so slowly that the war will be ended long before you have a sufficient number of troops in Europe to affect the result."

The submarine rolled a little in the Atlantic, though we had no very rough weather. In the North Sea the choppy seas seemed hardly to affect it; and under the surface there was no sensation of being in motion. The air inside the submarine when we were submerged on the last day for ten hours was becoming disagreeable. However, several tanks of oxygen were carried which Remy told me he would use in case of necessity. The water-tight doors between the different compartments were kept closed at all times after entering the North Sea. The officers and crew smoked in the conning tower or on deck, but nowhere else. The wardroom was about six feet wide and seven feet long. Here we ate at a small table, and in the lockers along the bulkhead the wardroom food was kept. Here also they installed hammock hooks and swung a hammock for me to sleep in alongside two bunks used by Kahn and one of the other officers.

Just forward of this room was a smaller compartment known as the captain's cabin, in which he had his desk and bunk — with scarcely room for either. Forward of this cabin was a sleeping compartment for the men, and forward of this was the forward torpedo room. I was never allowed in the torpedo rooms. Abaft the wardroom on the starboard side was a small cabin about four feet wide and six feet long occupied by the two other officers. Across the passage on the port side was the radio room. Abaft these two small compartments was the control room. Here there were always two men on watch. Abaft the control room was the other living compartment for the men. Here the food was cooked and the men ate their meals. Abaft this was the engine room and then the after torpedo room. The men slept in hammocks and on the deck. They were very dirty for there was no water to wash with. In the wardroom we had enough to wash our hands and faces every day, but that was all.

A little wine was carried for the officers, who also had eggs two or three times while I was on board. They had sausage at every meal, canned bread and lard, which they called marmalade and used on their bread. Remy told rde, however, that the people on the submarines were the only ones who had an unlimited amount of meat and the like. We had practically four meals every day; at 8:00 A. M., breakfast; at 12:00 o'clock noon, dinner; at 4:00 P. M., what they called "Kaffee," and at 8:00 P. M. supper, but practically every meal was the same, at least until we had the fresh mutton shot on North Bona Island. "Kaffee" at 4:00 P. M. apparently corresponded to our tea, but the sausage (or, as they call it, "Wurst") was placed on the table every meal.

After supper every night we played cards, sometimes bridge and sometimes a new game, with the secrets of which I was soon acquainted. Captain Remy tried in every way possible to make things pleasant for me, and when I asked an impossible question he invariably told me he did not think he ought to answer, so I have great confidence that what he did tell me was the truth.

The *U-90* and most of the other German submarines were out usually not more than five or six weeks, and then in port about three weeks. The service was not severe for Remy got leave as often as he cared to have it, and indeed it was deemed the height of good fortune by regular officers to be assigned to a submarine. The crew seemed happy and well fed. After making, I think, three round trips, they were entitled to the Iron Cross and to leave, which leave covered the duration of the stay of the submarine in port. They receive extra money and they get the best food in Germany; besides which, for every day that they submerge, both officers and men receive extra money. For all of these reasons it is a popular service. On this trip of the *U-90* she arrived back at Wilhelmshaven the thirty-third day after leaving Kiel.

On the trip we received news of German submarines being in American waters from the Radio Press. Remy was chagrined that he had not been allowed to go to America with the *U-90;* he told me he had previously requested it.

I was in my prison room on the *Preussen* two or three days. Twice I saw the Commanding Officer, who brought me a toothbrush and a comb. Remy came to see me twice before he went on leave and gave me cigarettes. He also changed into German money a $5 bill which I had found on my clothes. I had him get me some toothpaste and a few other toilet articles.

After the two visits from the Commanding Officer of the Preussen, I saw no more of him, and he apparently left my rationing and entertainment to my guards. Sometimes they brought me food and sometimes they didn't. Practically all the time I had only sour black bread which was almost impossible to eat, and some warm water colored with Ersatz coffee, which we afterwards found out was made of roasted acorns and barley.

A PRISONER IN GERMANY

Finally I was taken to the prison on shore, to what they call the Commandatur. I was escorted through the streets by a warrant officer wearing side arms and a guard of about four men. We landed from a launch and walked rapidly through the streets for about 45 minutes. At the Commandatur I was placed in a room which opened off a corridor. There was a guard in the corridor outside of my door; the door was kept locked at all times and there was another guard outside my window. The guards were armed with rifles which I noticed they kept loaded. Here they searched me and took my identification tag. They also took my gun and left me my binoculars. Up to this time I had had my gun. On board the submarine I cleaned, oiled and loaded it, keeping it on Remy's desk.

I was in the prison at Wilhelmshaven two days. A naval officer visited me twice and questioned me. My food was the same as it had been on the *Preussen*. At 5 o'clock the morning of the third day a young naval officer and two men came for me and took me to the station, where we boarded a train for Karlsruhe. It was then I realized how fortunate I was to have the $5 bill, for I had nothing to eat on the trip except a sandwich which the officer gave me from his lunch. However, at the station in Hanover he allowed me to buy a meal when he found that I had some money. We came by way of Hanover, Frankfort, Mannheil, to Karlsruhe. Near Wilhelmshaven there were large herds of Holstein cattle, apparently for the fleet. Those were about the only cattle in any numbers that I saw in all Germany.

When we arrived at Karlsruhe, I was taken to what prisoners call the "Listening Hotel" and there turned over to the Army authorities. The procedure in this hotel is as follows: An officer is placed in a room alone; the doors and windows are locked; he cannot see outside, and he is in communication with no one. After a day of this he is placed with an officer who speaks the same language. In this room there are dictaphones hidden under tables, in chandeliers and in similar places. In this way the Germans try to get information of military value.

My second day at this hotel I was placed with eight Frenchmen in another room, and on the third day in a room with three British officers. While we were there three dictaphones were found by the officers, and little time was lost in tearing them out and destroying them.

On the fourth day I was sent to the officers' camp in the Zoölogical Gardens at Karlsruhe. Here I found about 20 Italians, 10 Serbs, 100 French and 50 British officers. Among this number were one French Naval officer by the name of Domiani and a British Warrant officer, who had also been prisoners on board U-boats. From them I got some valuable data which checked up with the information I had picked up on board the *U-90.* This information I considered of importance to enable the Allies to locate and attack enemy submarines and I determined to escape.

I was the only American at Karlsruhe, but the British and French treated me as one of themselves, and when they heard I intended to escape they provided me with maps, a compass, money and food. For two weeks I worked on plans for my escape. Two plans failed; the third (in which I was associated with some British and French officers) failed when a letter written by one of the French officers to a woman in Karlsruhe fell into the hands of the Commandant of the camp. The aviator had been in Karlsruhe before the war and had many friends there. Through one of the guards he had communicated with one of these, a woman, and she had assisted in our plans. When the Commandant found the letter he suspected a big camp delivery, so Berlin was notified immediately.

The following day orders came from Berlin to clear the camp of all officers. In the forenoon all the British left except the aviators; these were followed in the afternoon by all the aviators and the French officers. There then remained only a few Italians, some Serbian officers, two British generals and myself.

I found the generals real live wires, and with one of them I made plans for a fresh attempt. We could not try that night and anyway it looked as if we were to be left there indefinitely and so could wait for, a better opportunity. The following morning at 6 o'clock one of the interpreters woke me and told me to be ready to leave the camp in half an hour. I dressed and hid my compass and maps as best I could in the short time, and passed through my search without anything being found.

Upon entering and leaving a camp each officer is searched thoroughly. If any suspicion is aroused the officer is required to take off all his clothes and each garment is separately inspected, kneaded to see if the rustle of paper can be heard, and finally the hems are ripped open, gold stripes and insignia cut off to see if a map or some other contraband is secreted within. Even the soles and heels of the shoes are cut off in their search — as happened in my case.

I had no regret in leaving that camp for I felt that I could not be much worse off, and I might possibly find conditions better at the next camp. Besides, we considered a journey the best time for attempting to escape. At Karlsruhe we had no breakfast. At noon we had soup made out of leaves, and a plate of black potatoes or horse carrots, or something similar. At night the same kind of soup again, and that was all, except the 240 grammes of black bread which we received every day.

At Karlsruhe I spent about three weeks and in all that time the soup was never changed. It was absolutely tasteless. It was hardly possible to exist on that ration, but the British and French Red Cross Committees had enough food to considerably ameliorate conditions. The French Committee had orders from France to take care of Americans, and while they had very few supplies, I was given what they did have in like manner to their own countrymen.

The morning I left Karlsruhe, I noticed that all the Serbians and about 20 Frenchmen who had come in the night before, were also leaving camp. They were guarded by four sentries. I had two. I was marched through the town to the station and on to the train. The guards then told me we were bound for Villingen and would get there about 3:00 P. M. I saw a time table and planned to jump from the train at the first opportunity, but preferably as far south as possible in order not to have so far to walk to reach the Swiss frontier. But never once had I the least opportunity of breaking from the guards. They sat on either side of me with their guns (which were loaded) pointed at me at all times. Finally we were only a few miles from Villingen, the train had already reached and passed the crest of the mountains and was on the down grade making good speed. I knew it had to be now or not at all. So watching my chance I caught one guard half dozing and the other with his head turned in the other direction, and jumping past them I dove for the window. It was very small, probably 18x24 inches. On the outside of the car there was nothing to land on so I simply fell to the ground. Just as I disappeared, the guards who had been wondering what it all was about, jumped to their feet with a shout and pulled the bell cord. The train came to a stop about 300 yards farther on.

In the meantime I had landed on the second railway track. The ties were of steel and in falling I struck my head on one and was stunned for a few seconds. But the injury that did the damage was to my knees which struck another tie and were cut so badly that I could not bend them. I struggled to my feet and tried to shuffle off towards the hills and forest a few hundred yards away. But by this time the guards were out of the train and firing at me. I kept on going as long as I could, and then turned around and found that the guards were only 75 yards away, so I held up my hands as a sign that I surrendered. One of the guards had just fired. The shot passed between my hat and shoulder, and had they continued firing they must surely have hit me. When I turned they were on me in a few seconds. The first guard beat me with the butt of his rifle as I half lay and half sat on the side of the hill. I remember rolling down hill, gaining additional impetus from their boots. They kicked me until I got up, and when I was up they knocked me down again with their guns. I noticed many people working in the fields who came over to look on. Finally in knocking me down the seventh or eighth time one of the guards struck me and his gun broke in two at the small of the stock. Villingen was about five miles away. They marched me down the road at as near double time as I could make shuffling along. They were beating and kicking me continuously. We finally arrived at the prison camp and I collapsed on the guardhouse porch. I was greeted by the Commandant, a porkish looking individual and typically Prussian, who bellowed at me in German that if I attempted to escape again I would be shot. An interpreter told me what he said. They sent for the German doctor and he bandaged me from head to foot with the paper bandages they use.

Then I was put on a bed in one of the guardhouse cells. For three days I could not move and the vermin that infected the place made it almost unbearable. Later, when I had recuperated enough to move my arms and upper body, I was able to keep most of the vermin away while I was awake. My body was covered with large red eruptions, for the German fleas are as poisonous as German propaganda.

About my sixth day in the cell, I was given a court-martial, or at least I would call it such. There were three officers, and after questioning me they decided that I should be given two weeks' solitary confinement in my cell. They never stopped the food and books that the American officers sent in to me, so I was not so badly off as I might have been. When I came out of the cell, however, I weighed only 120 pounds — I had lost 30.

Thereupon I began to consider fresh plans for escape. Thanks to Red Cross food, I built up and got myself in good physical trim. Three plans failed due to treachery. There must have been some spies among the Russian officers, who gave our plans to the Germans. We were very much handicapped there because all the orderlies were Russian and the Russian officers themselves included every variety from the regulars captured in 1914 to some Bolsheviki. We could trust no one. Our own officers included more than 25 combatants, about 20 doctors and five merchant officers taken by the raider *Wolf.*

At Villingen the food was practically the same as at Karlsruhe, probably a little better. At least we did not notice that it was so bad because we seldom ate it, having instead our regular parcels from the Red Cross.

The Germans had finally decided to make Villingen an exclusively American camp. On October 7th all the Russian officers were to be shifted to the north of Germany. We knew that meant a thorough search for the following day. Once before we had undergone a search but fortunately the Germans were deceived by the exemplary conduct of the men in my barracks, and passed us by. I had a complete set of tools, over 100 large screws taken from all the doors in the camp, and four long chains made out of wire, which, a few days previously, had enclosed the tennis court. All these things were necessary in almost any plan of escape that we might devise, and I could not afford to lose them. In the other barracks they found several compasses, maps and other contraband. On one aviator they found a map sewed inside the double seat of his trousers. This cost him six days' solitary confinement. But we had suffered one disaster in this search: that was the loss of our material for ladder building which we had prepared out of bedslats after prolonged efforts.

THE ESCAPE

On Sunday, October 6th, the day before the Russians were to leave camp, I called a meeting in my barracks of the 12 other officers whom I knew were interested in getting away. I insisted that we go that night. Our plan was to try and go over or cut through the fences in different parts of the yard simultaneously. We divided up into four teams. I had the first team, consisting of two aviators and myself; Major Brown the second team, consisting of one of the aviators and two infantry officers; Lieutenant Willis of the Lafayette Escadrille the third team, consisting of three other aviators; the fourth team was composed of two aviators who decided to go at the last minute.

The defensive works of the camp consisted first of the barred windows in the barracks, which ran along parallel to the outer fences; then a ditch filled with barbed wire and surmounted by a four-foot barbed wire fence. This was about eight feet outside the line of barracks. About seven feet outside the ditch was the last artificial defense — a barbed wire fence about eight or ten feet high with top wires curved inward out of the vertical plane of the rest of the fence. This was to prevent any one from climbing up and over, which would have been simple with a fence straight up and down. Outside the outer fence was a line of sentries about one for every 30 yards, and inside the yard there were two sentries who patrolled at their discretion.

The plan of the first team was to cut the iron grating of the window in my barracks and launch a bridge through the opening out to the top of the outer barbed wire fence. "We were to then crawl along the bridge and drop down outside the wire. The second team had wire cutters and were to cut through the outer wire. The third team were to go out of the main gate with the guard off duty when it rushed out in pursuit of the other teams. The fourth team were to build a small ladder and climb over the outer fence.

At 10:30 the barracks lights were turned out as usual. Shortly afterwards the signal was given and a team consisting of doctors threw the chains and short circuited all the lighting circuits in the camp.

I have never been able to find out how the other teams fared, except to know that Willis of the third team and one of the fourth team got out of the camp. My team was more successful. The night before one of the officers and I stole out to the tennis court and brought into my barracks the two long wooden battens used as markers. We hid them under the beds. They were about 2 ½ inches wide, one inch thick and were 18 feet long. I had had my eye on them for a long time because they were the only things in the camp to reach from the window ledge to the outer barbed wire fence. They were very light and of course would not hold any weight, but I had a plan to remedy that. Two Army officers who did not care to go were to launch the bridge through the window to the outer fence, leaving the three-foot overlap on the inboard side. When we crawled over the bridge they would then put their weight on the ends that overlapped and this would neutralize the great bending moment at the middle of the span.

I had stolen Red Cross food boxes and with the boards from these I made little flats which when screwed to the long battens (nailing would have attracted the guards) would make a very passable bridge. In the afternoon one of my team and I cut and filed the grating in my window. It had to be done when the guards were at the end of their beats outside, but we finally finished by dark. After last muster at 7:00 P. M. we began on the bridge and finished it by 10:00 o 'clock. I then blackened it with shoe blacking so it would not appear white in the darkness.

As the lights went out the bridge was thrown across and the smallest in the team of three crawled out. I was second and the heaviest man third. When the bridge struck the outer fence, the nearest guards ran to the spot singing out: "Halt! Halt!" As the first man reached the end of the bridge and dropped to the ground outside, I was beside him before he could straighten up and coaching him I dashed past the guards, who were then within a few feet of us preparing to fire. As we passed them they fired and the flash of the gun on my right almost scorched my hair. Then I heard the third man jump to the ground. We continued to run directly away from the camp and the whole side opened fire. Although the bullets were singing all around us, we were not hit. By our thus drawing fire, the other teams had a fine opportunity to cut their way out.

A few minutes later the guard of about 40 men sleeping in the guardhouse rushed out of the main gate in answer to the firing, and Willis came out with them, was fired on, but finally kept his rendezvous with me about two miles away. Knowing that in a few minutes the battalion of at least 300 men, together with hounds, would be on our trail, we headed across country and put several miles between us and the camp. We continued thus for six days and nights, walking mostly in the night time, never on roads and bridges, which are patrolled, but through the rivers, fields and mountains, and finally on the seventh night we came to the Rhine.

We had travelled about 120 miles, although the distance as the crow flies is perhaps only about 40 miles. We had a little food in our pockets, but lived mostly on the raw vegetables in the fields. When we came to the Rhine we spent about four hours trying to get past the sentries, and finally had to crawl the last half mile on our hands and knees down the bed of a mountain creek.

About 2:00 A. M., Sunday, October 13, we were crouching in the water at the mouth of this creek where it flows into the Rhine. The hardest fight was still before us. In whispers we discussed the next move and then took off most of our clothes. As we stepped farther out, the current caught us and swept us away. The stream at this point is 200 meters wide and has a current of 12 kilometers an hour. The water was like ice, but when I had been carried to the center of the stream I couldn't get out. After fighting for ten minutes, I made one last effort and managed to get past the worst of the center, and then just as the last of my strength had gone my feet touched the rocks.

I was then in Switzerland. After a rest I crawled up the bank and in a few minutes found a house, where I was taken in and put to bed. The next morning I was turned over to the gendarmes. They had also located Willis in a house about three miles further down, where he found himself after his swim.

The Swiss were elated when they heard we were Americans. They took us to Berne and turned us over to the American Legation on October 15th, where we were provided with passports. While there we were interviewed by the American Commission for the exchange of prisoners of war. We borrowed money from the American Red Cross and proceeded to Paris and there awaited orders from October 18th to 21st. I was ordered to London, where I had asked to be sent, arrived October 23rd, and reported to Vice Admiral Sims, to whom I gave my information in the form of a detailed report. The British Admiralty kept me for three days and it was November 2nd before I left England, being then ordered to report to the Bureau of Navigation, Washington, B, C, where I arrived November 11, 1918.

Appendices

Table A - Organization, Cruiser and Transport Force, United States Atlantic Fleet, July 1, 1916

Rear Admiral ALBERT GLEAVES, Commander

CRUISER FORCE

SQUADRON ONE (Rear Admiral Albert Gleaves)

Division ONE	Division TWO	Division THREE	Special Duty
SEATTLE (Flag)	SOUTH DAKOTA	COLUMBIA	Division
NORTH CAROLINA	PUEBLO	MINNEAPOLIS	NIAGARA
MONTANA	FREDERICK	DE KALB	DUBUQUE
HUNTINGTON	SAN DIEGO	VON STEUBEN	

SQUADRON TWO (Rear Admiral Marbury Johnston)

Division FOUR (Rear Admiral H. P. Jones)	Division FIVE	Division SIX
SIALIA (Flag)	ISIS (Flag)	ALBANY
CHARLESTON	DENVER	NEW ORLEANS
ST. LOUIS	GALVESTON	TACOMA
ROCHESTER	CLEVELAND	CHATTANOOGA
OLYMPIA	DES MOINES	

FRENCH MEN-OF-WAR OPERATING WITH CRUISER FORCE
(Rear Admiral Grout)

GLOIRE (Flagship)   MARSEILLAISE   DU PETIT THOUARS

TRANSPORT FORCE

NEW YORK DIVISION (Rear Admiral Albert Gleaves)

AGAMEMNON	LENAPE	ORIZABA
AMERICA	LEVIATHAN	PLATTSBURG
CALAMARES	LOUISVILLE	PRESIDENT GRANT
FINLAND	MALLORY	PRINCESS MATOIKA
GEORGE WASHINGTON	MANCHURIA	RIJNDAM
GREAT NORTHERN	MATSONIA	SIBONEY
HANCOCK	MAUI	SIERRA
HARRISBURG	MONGOLIA	ST. PAUL
HENDERSON	MOUNT VERNON	WILHELMINA
KROONLAND	NORTHERN PACIFIC	

NEWPORT NEWS DIVISION (Rear Admiral H. P. Jones)

AEOLUS	MARTHA WASHINGTON	POWHATAN
ANTIGONE	MERCURY	SUSQUEHANNA
HURON	PASTORES	TENADORES
MADAWASKA	POCAHONTAS	ZEELANDIA

FOREIGN VESSELS OPERATING WITH TRANSPORT FORCE
NEWPORT NEWS DIVISION

AMERICA	DANTE ALIGHIERI	FRANCE	PATRIA
CASERTA	DUCA DEGLI ABRUZZI	KURSK	RE D'ITALIA
CZAR	DUCA D'AOSTA	LUTETIA	SOBRAL
CZARITZA			

Table B

Report by Months of Transport and Escort Duty Performed by U. S. and Foreign Navies up to Signing of the Armistice

1917	Carried by U. S. N. Transports	No. of U. S. N. Transports Sailed	Carried by British Ships	No. of British Ships Sailed	Carried by British Leased Italian Ships	No. of British Leased Italian Ships Sailed	Carried by Other U. S. Ships	No. of Other U. S. Ships Sailed
May	0	0	508	2	0	0	1035	3
June	8855	9	1080	1	0	0	5156	8
July	5281	8	7299	6	0	0	0	0
Aug.	4310	6	11890	7	0	0	1109	2
Sept.	13917	15	19671	12	0	0	0	0
Oct.	25098	14	13013	9	0	0	0	0
Nov.	9988	9	10669	7	0	0	1235	2
Dec.	37445	16	11370	9	0	0	0	0
1918								
Jan.	25662	16	20514	9	0	0	0	0
Feb.	39977	17	9259	4	0	0	0	0
Mar.	58278	26	27626	14	0	0	1	1
Apr.	67553	27	47362	20	2626	2	737	11
May	96273	33	133795	75	12127	6	3288	22
June	115256	36	140172	70	14465	7	6003	11
July	108445	33	175526	89	11502	7	4020	13
Aug.	116401	36	137745	74	9376	6	8495	15
Sept.	107025	35	134576	69	7052	4	5511	18
Oct.	72092	43	94214	57	11098	7	4709	17
To Nov. 11	1191	9	10698	12	0	0	235	3
Grand Total	911047	388	1006987	546	68246	39	41534	126

1917	Carried by Other Ships, French, Italian, etc.	No. of Other Ships Sailed	Carried by U. S. Navy Transports and by Other U. S. Ships	No. of U. S. N. Transports and Other U. S. Ships Sailed	Total Transported by All Ships	Total Ships Sailed	% Carried by U. S. N. Transports	% Carried by British	% Carried by British Leased Italian Ships
May	0	0	1035	3	1543	5	67	33	0
June	0	0	14011	17	15091	18	59	6.5	0
July	296	1	5281	8	12776	15	41	57	0
Aug.	2094	2	5419	8	19403	17	22	61	0
Sept.	0	0	13917	15	33588	27	41	59	0
Oct.	1916	1	25098	14	40027	24	62.5	32.5	0
Nov.	1830	1	11223	11	23722	19	41.5	46	0
Dec.	0	0	37445	16	48815	25	77	23	0
1918									
Jan.	1879	1	25662	16	48055	26	53	42.5	0
Feb.	3	1	39977	17	49239	22	81.5	18.5	0
Mar.	1895	4	56279	27	85710	45	65	·33	0
Apr.	1794	5	68290	38	120072	63	56	39.5	2
May	2231	5	99561	55	247714	141	39	53.5	5
June	4538	4	121259	47	280434	128	41.25	50	5
July	11866	5	112465	46	311359	147	35	56.5	3.5
Aug.	14358	9	124896	51	286375	140	41	48	3
Sept.	5506	3	112536	53	259670	129	41	52	3
Oct.	1950	3	76801	60	184063	127	39	51	6
To Nov. 11	0	0	1426	12	12124	24	10	88	0
Grand Total	52066	43	952581	514	2079880	1142	43.75	48.25	3

1917	% Carried by Other U. S. Ships	% Carried by Other Ships, French, Italian, etc.	% Carried by U.S.N. Transports and Other U. S. Ships	Under U. S. Escort	Under British Escort	Under French Escort	% Under U. S. Escort	% Under British Escort	% Under French Escort
May	0	0	67	258	1285	0	17	83	0
June	34.5	0	93.5	15032	59	0	99	1	0
July	0	2	41	10063	2566	247	78.5	20	1.5
Aug.	6	11	28	12259	4129	3015	63	21	16
Sept.	0	0	41	17432	12898	3258	51.5	39	9.5
Oct.	0	5	62.5	36893	3134	0	92.5	7.5	0
Nov.	4.5	8	46	13246	10476	0	56.5	43.5	0
Dec.	0	0	77	42783	6092	0	87.5	12.5	0
1918									
Jan.	0	4.5	53	35827	12228	0	75	25	0
Feb.	0	0	81.3	48795	444	0	99	1	0
Mar.	0	2	65	73995	12615	0	85	15	0
Apr.	1	1.5	57	91308	28764	0	75.5	24.5	0
May	1.5	1	40.5	220463	26652	599	88.5	11	.5
June	2.25	1.5	43.5	244631	30912	4891	87.5	11	1.5
July	1	4	36	258332	46329	6698	83	15	2
Aug.	3	5	44	237920	22572	25883	83	8	9
Sept.	2	2	43	224298	20081	14691	86	8	6
Oct.	3	1	42	130274	31454	2335	70.5	28.5	1
To Nov. 11	2	0	12	7431	4673	0	61.75	38.25	0
Grand Total	2.5	2.5	46.25	1720360	297903	61617	82.75	14.125	3.125

Table C

REPORT BY MONTHS OF TRANSPORT DUTY PERFORMED BY U. S. NAVY AND ALL OTHER SHIPS, U. S. AND FOREIGN, IN RETURNING TROOPS AND OTHER PASSENGERS TO U. S. PRIOR TO SIGNING OF ARMISTICE

Month	Carried by Cruiser and Transport Force	Carried by All Other Ships U. S. and Foreign	Total Carried All Ships	% Carried by Cruiser and Transport Force	% Carried by All Other Ships
1917					
May					
June					
July					
August					
September					
October	41	6	47	87.3	12.7
November	37		37	100	
December					
1918					
January	66	1	67	98.6	1.4
February	274	86	360	76	24
March	402	86	488	82.3	17.7
April	508	46	554	91.7	8.3
May	544	39	583	93.3	6.7
June	368	101	469	78.4	21.6
July	946	23	969	97.6	2.4
August	1920	67	1987	96.6	3.4
September	1710	56	1766	97	3
October	3436	306	3742	91.8	8.2
To Nov. 11	959	183	1142	84	16
Total	11211	1000	12211	91.8	8.2

Table D

Month	Carried by Cruiser and Transport Force	Carried by All Other Ships U. S. and Foreign	Total Carried All Ships	% Carried by Cruiser and Transport Force	% Carried by All Other Ships
1918					
From Nov. 11	7689	508	8197	93.9	6.1
December	47228	22861	70089	67.2	32.8
1919					
January	97039	23097	120136	80.8	19.2
February	96368	44463	140831	68.3	37.7
March	165312	42049	207361	79.7	20.3
April	243697	30806	274503	88.8	11.2
May	278600	34610	313210	89.0	11.0
June	314167	26779	340946	92.0	8.0
July	268049	27162	295211	90.8	9.2
August	112694	2127	114821	98.0	2.0
September	44890	2961	47851	93.8	6.2
Total	1675733	257423	1933156	86.7	13.3

Table E – Record of Ships

The following Naval Transports were used in transporting troops to and from France employed in transporting

No.	Name of Ship / Date Placed in Commission or Attached to Force / Type of Vessel	Displacement	Original Troop Carrying Capacity Including Officers	Maximum Troop Carrying Capacity Including Officers	Voyages Made Prior to Signing of Armistice		
					Number of Turn-arounds Made	Total Number of All Passengers Carried to Europe	Total Number of Passengers Returned from Europe
1	Aeolus / Aug. 4–17, Ex-German	22000	2800	3500	8	24770	400
2	Agamemnon / Aug. 21–17, Ex-German	30000	3400	5800	10	36097	214
3	America / Aug. 6–17, Ex-German	41500	4000	7000	9	39768	168
4	Antigone / Sept. 5–17, Ex-German	15000	2000	3500	8	16526	101
5	Calamares / Apr. 9–18, Am-Passenger	10000	1400	2200	5	7657	0
6	Covington / July 28–17, Ex-German	41500	3400	4100	6	21628	0
7	De Kaib (Aux. Cruiser) / May 12–17, Ex-German	14280	800	1600	11	11334	48
8	Finland / Apr. 26–18, Am-Passenger	22000	3500	3800	5	12654	16
9	Geo. Washington / Sept. 6–17, Ex-German	39435	5600	6500	9	48373	484
10	Great Northern / Nov. 1–17, Am-Passenger	14000	2800	3300	10	28248	677
11	Hancock / Marine Transport	10000	1000	1000	2	1438	0
12	Harrisburg / May 29–18, Am-Passenger	15000	2100	2600	4	9855	0
13	Henderson / May 24–17, Marine Transp.	10000	1800	2500	10	16352	112
14	Huron / July 25–17, Ex-German	15000	2300	3400	8	20871	67
15	K. der Nederlanden / Apr. 4–18, Dutch Chart'r'd	13600	2200	2200	3	6283	0
16	Kroonland / Apr. 25–18, Am-Passenger	22000	3300	3800	5	14125	77
17	Lenape / Apr. 24–17, Am-Passenger	7000	1200	1900	6	8975	0

No.	Name of Ship Date Placed in Commission or Attached to Force Type of Vessel	Displacement	Original Troop Carrying Capacity Including Officers	Maximum Troop Carrying Capacity Including Officers	Voyages Made Prior to Signing of Armistice		
					Number of Turn-arounds Made	Total Number of All Passengers Carried to Europe	Total Number of Passengers Returned from Europe
18	Leviathan July 25-17, Ex-German	60000	9000	12000	10	96804	650
19	Louisville Apr. 27-18, Am-Passenger	14000	2300	2500	4	9247	14
20	Madawaska Aug. 27-17, Ex-German	15000	2000	2800	9	17931	21
21	H. R. Mallory Apr. 17-18, Am-Passenger	11000	1800	2000	6	9756	0
22	Manchuria Apr. 25-18, Am-Passenger	26500	3500	4800	4	14491	16
23	Martha Washington Jan. 2-18, Ex-German	14500	2800	3400	8	22311	185
24	Matsonia March 1-18, Am-Passenger	17000	2300	3400	6	13329	10
25	Maui March 6-18, Am-Passenger	17500	3500	3800	4	11042	11
26	Mercury Aug. 3-17, Ex-German	16000	2900	3200	7	18542	20
27	Mongolia May 8-18, Am-Passenger	20695	3700	4700	5	19013	24
28	Mount Vernon July 28-17, Ex-German	32130	3100	5800	9	33692	86
29	Northern Pacific Nov. 1-17, Am-Passenger	12500	2400	2800	9	20711	38
30	Orizaba May 27-18, Am-Passenger	13000	3100	4100	6	15712	16
31	Pastores May 6-18, Am-Passenger	13900	1600	2100	6	9928	99
32	Plattsburg May 25-18, Am-Passenger	10000	2300	2600	4	8776	411
33	Pocahontas July 25-17, Ex-German	14500	2400	2900	9	20503	221
34	Powhatan Aug. 16-17, Ex-German	17000	1800	3100	7	14613	46
35	Pres. Grant Aug. 2-17, Ex-German	33000	4800	5900	8	39974	0
36	Pres. Lincoln July 25-17, Ex-German	29000	3800	4700	5	20143	0
37	Princess Matoika May 27-18, Ex-German	17500	3500	3900	6	21216	206
38	Rijndam May 1-18, Dutch Chart'r'd	22070	3100	3700	6	17913	439

No.	Name of Ship Date Placed in Commission or Attached to Force Type of Vessel	Displacement	Original Troop Carrying Capacity Including Officers	Maximum Troop Carrying Capacity Including Officers	Voyages Made Prior to Signing of Armistice		
					Number of Turn-arounds made	Total Number of All Passengers carried to Europe	Total Number of Passengers returned from Europe
39	Siboney Apr. 8-18, Am-Passenger	11250	3100	4000	7	20219	11
40	Sierra July 1-18, Am-Passenger	10000	1500	1700	1	1712	0
41	Susquehanna Sept. 5-17, Ex-German	16950	2200	3300	8	18345	0
42	Tenadores Apr. 17-18, Am-Passenger	10000	1200	1200	13	15698	8
43	Von Steuben June 9-17, Ex-German	22000	1200	2900	9	14347	21
44	Wilhelmina Jan. 26-18, Am-Passenger	13500	1800	2100	6	11053	90
45	Zeelandia Apr. 3-18, Dutch Chart'r'd	12950	1800	3000	5	8349	8
	Total	879860	122100	161100	306	870324	5051

Table – Record of Battleships and Cruisers

No.	Name of Ship / Date Attached to Force or Readiness for Transporting Troops / Type of Vessel	Displacement	Original Troop Carrying Capacity Including Officers	Maximum Troop Carrying Capacity Including Officers	Voyages Made Prior to Signing of Armistice		
					Number of Turn-arounds Made	Total Number of All Passengers Carried to Europe	Total Number of Passengers Returned from Europe
46	Charleston Jan. 17–17, Cruiser	10839	1700	1700	0	0	0
47	Connecticut Dec. 25–18, Battleship	16000	1000	1300	0	0	0
48	Frederick Jan. 2–19, Cruiser	13720	1600	1700	0	0	0
49	Georgia Dec. 10–18, Battleship	14948	900	1400	0	0	0
50	Huntington Dec. 14–18, Cruiser	13720	1700	2000	0	0	0

No.	Name of Ship / Date Attached to Force or Readiness for Transporting Troops / Type of Vessel	Displacement	Original Troop Carrying Capacity Including Officers	Maximum Troop Carrying Capacity Including Officers	Voyages Made Prior to Signing of Armistice		
					Number of Turn-arounds made	Total Number of All Passengers carried to Europe	Total Number of Passengers returned from Europe
51	Kansas Dec. 10–18, Battleship	16000	1600	1900	0	0	0
52	Louisiana Dec. 21–18, Battleship	15000	900	1400	0	0	0
53	Michigan Dec. 21–18, Battleship	16000	1000	1000	0	0	0
54	Minnesota Feb. 25–19, Battleship	16000	1200	1400	0	0	0
55	Missouri Mar. 6–19, Battleship	12240	700	1000	0	0	0
56	Montana Jan. 12–19, Cruiser	14375	1300	1500	0	0	0
57	Nebraska Dec. 28–18, Battleship	16325	1000	1200	0	0	0
58	New Hampshire Dec. 21–18, Battleship	18664	1000	1300	0	0	0
59	New Jersey Dec. 28–18, Battleship	14046	1000	1400	0	0	0
60	North Carolina Dec. 23–18, Cruiser	14372	1200	1500	0	0	0
61	Ohio Feb. 4–19, Battleship	14150	700	700	0	0	0
62	Pueblo Jan. 18–19, Cruiser	13300	1550	1800	0	0	0
63	Rhode Island Dec. 17–18, Battleship	14948	900	1100	0	0	0
64	Rochester Jan. 14–19, Cruiser	8150	300	900	0	0	0
65	Seattle Dec. 21–18, Cruiser	15000	1500	1600	0	0	0
66	South Carolina Feb. 18–19, Battleship	16000	1100	1400	0	0	0
67	South Dakota Dec. 21–19, Cruiser	14000	1600	1800	0	0	0
68	St. Louis Dec. 18–18, Cruiser	9700	1300	1400	0	0	0
69	Vermont Jan. 7–19, Battleship	16000	1000	1200	0	0	0
70	Virginia Dec. 17–18, Battleship	11980	900	1400	0	0	0
	Total	358477	28650	34400	0	0	0

Table – Merchant Ships Converted into Troop Transports

No.	Name of Ship Date Placed in Commission or Attached to Force Type of Vessel	Displacement	Original Troop Carrying Capacity Including Officers	Maximum Troop Carrying Capacity Including Officers	Voyages Made Prior to Signing of Armistice		
					Number of Turn-arounds Made	Total Number of All Passengers Carried to Europe	Total Number of Passengers Returned from Europe
71	Alaskan Dec. 12–18, Am-Cargo	8000	2100	2300	0	0	0
72	Amphion Apr. 12-19, Ex-Ger'n Cargo	15530	2400	2500	0	0	0
73	Ancon Mar. 28-19, Am-Cargo	20000	3000	3100	0	0	0
74	Arcadia Jan. 20-19, Ex-Ger'n Cargo	7900	1000	1100	0	0	0
75	Arizonian Aug. 14–18, Am-Cargo	18500	2500	2600	0	0	0
76	Artemis Apr. 8-19, Ex-Ger'n Cargo	12540	3800	4000	0	0	0
77	Black Arrow Jan. 27-19, Ex-Ger'n Cargo	12200	1500	1600	0	0	0
78	Buford Jan. 15-19, Army Trans.	10000	1000	1200	0	0	0
79	Callao Apr. 26-19, Ex-Ger'n Cargo	13164	2400	2400	0	0	0
80	Canandaigua Mar. 2-19, Am-Cargo	7610	1400	1400	0	0	0
81	Cananocius Mar. 8-19, Am-Cargo	7500	1400	1400	0	0	0
82	Cape May Jan. 29-19, Am-Cargo	10350	1800	1900	0	0	0
83	Comfort Mar. 18-18, U. S. N. Hosp. Ship	10000	300	400	0	0	0
84	Dakotan Jan. 29-19, Am-Cargo	14375	1500	2000	0	0	0
85	Eddelyn July 18-19, Am-Cargo	12500	985	985	0	0	0
86	El Sol Aug. 5-18, Am-Cargo	10000	1800	1800	0	0	0
87	El Oriente April 11-19, Am-Cargo	11000	2000	2000	0	0	0
88	Etten May 1-19, Ex-Ger'n Cargo	6900	1500	1800	0	0	0
89	Eurana Sept. 13-18, Am-Cargo	15250	1800	1800	0	0	0

No.	Name of Ship Date Placed in Commission or Attached to Force Type of Vessel	Displacement	Original Troop Carrying Capacity Including Officers	Maximum Troop Carrying Capacity Including Officers	Voyages Made Prior to Signing of Armistice		
					Number of Turn-arounds Made	Total Number of All Passengers Carried to Europe	Total Number of Passengers Returned from Europe
90	Floridian Jan. 28-19, Am-Cargo	9800	1700	1800	0	0	0
91	Freedom Jan. 24-19, Ex-Ger'n Cargo	11175	1600	1700	0	0	0
92	Gen. Goethals Mar. 10-19, Ex-Ger'n Cargo	7700	1400	1400	0	0	0
93	Gen. Gorgas Mar. 8-19, Ex-Ger'n Cargo	5300	1000	1100	0	0	0
94	Housatonic Feb. 27-19, Am-Cargo	7522	1400	1400	0	0	0
95	Iowan Dec. 23-17, Am-Cargo	13912	1800	2000	0	0	0
96	Kentuckian Jan. 29-19, Am-Cargo	14405	1900	1900	0	0	0
97	Lancaster June 19-19, Am-Cargo	11500	2000	2000	0	0	0
98	Liberator July 28-18, Am-Cargo	12000	2500	2500	0	0	0
99	E. F. Luckenbach June 11-18, Am-Cargo	20000	2200	2300	0	0	0
100	Edward Luckenbach Dec. 30-19, Am-Cargo	5600	2200	2400	0	0	0
101	F. J. Luckenbach Feb. 22-19, Am-Cargo	12000	2400	2400	0	0	0
102	Julia Luckenbach Jan. 17-19, Am-Cargo	18390	2700	2700	0	0	0
103	Katrina Luckenbach May 18-18, Am-Cargo	15000	2250	2250	0	0	0
104	K. I. Luckenbach Aug. 9-18, Am-Cargo	16000	2300	2400	0	0	0
105	W. A. Luckenbach Dec. 14-18, Am-Cargo	17170	2400	2600	0	0	0
106	Marica June 9-19, Am-Cargo	17700	2000	2000	0	0	0
107	Mercy Jan. 24-18, U. S. N. Hosp. Ship	10100	400	400	0	0	0
108	Mexican Dec. 13-18, Am-Cargo	18200	2500	2500	0	0	0

No.	Name of Ship / Date Placed in Commission or Attached to Force / Type of Vessel	Displacement	Original Troop Carrying Capacity Including Officers	Maximum Troop Carrying Capacity Including Officers	Voyages Made Prior to Signing of Armistice		
					Number of Turn-arounds Made	Total Number of All Passengers Carried to Europe	Total Number of Passengers Returned from Europe
109	Minnesotan Jan. 8-19, Am-Cargo	14375	2000	2000	0	0	0
110	Montpelier Mar. 12-19, Ex-Ger'n Cargo	16430	2100	2300	0	0	0
111	Nansemond Jan. 20-19, Ex-Ger'n Cargo	27000	4900	5800	0	0	0
112	Ohioan Aug. 7-18, Am-Cargo	13345	1600	1900	0	0	0
113	Otsego Feb. 8-19, Ex-Ger'n Cargo	8750	1000	1000	0	0	0
114	Panaman Aug. 12-18, Am-Cargo	14495	2100	2200	0	0	0
115	Paysandu Jan 29-19, Ex-Ger'n Cargo	5750	1400	1400	0	0	0
116	Peerless Mar. 28-19, Am-Cargo	4214	2300	2300	0	0	0
117	Philippines May 1-19, Ex-Ger'n Cargo	18650	4000	4000	0	0	0
118	Radnor Mar. 6-19, Am-Cargo	14000	2000	2000	0	0	0
119	Roanoke Mar. 19-19, Am-Cargo	6500	1400	1400	0	0	0
120	Santa Ana Feb. 11-19, Am-Cargo	9000	1400	1700	0	0	0
121	Santa Barbara Feb. 21-19, Am-Cargo	9400	1600	1600	0	0	0
122	Santa Cecilia May 20-18, Am-Cargo	11000	2000	2000	0	0	0
123	Santa Clara Jan. 18-19, Am-Cargo	13000	1600	1800	0	0	0
124	Santa Elena Apr. 26-19, Ex-Ger'n Cargo	13000	900	900	0	0	0
125	Santa Elisa June 15-19, Am-Cargo	9345	1400	1400	0	0	0
126	Santa Leonora July 7-19, Am-Cargo	9345	1400	1400	0	0	0
127	Santa Malta Feb. 19-19, Am-Cargo	13340	1700	1700	0	0	0

No.	Name of Ship / Date Placed in Commission or Attached to Force / Type of Vessel	Displacement	Original Troop Carrying Capacity Including Officers	Maximum Troop Carrying Capacity Including Officers	Voyages Made Prior to Signing of Armistice		
					Number of Turn-arounds Made	Total Number of All Passengers Carried to Europe	Total Number of Passengers Returned from Europe
128	Santa Olivia / Dec. 20-18, Am-Cargo	9400	1900	1900	0	0	0
129	Santa Paula / Jan. 29-19, Am-Cargo	13500	2100	2200	0	0	0
130	Santa Rosa / Mar. 10-19, Am-Cargo	10000	2100	2100	0	0	0
131	Santa Teresa / Nov. 8-18, Am-Cargo	6900	1800	2000	0	0	0
132	Scranton / Feb. 5-19, Am-Cargo	14000	1900	1900	0	0	0
133	Shoshone / Feb. 19-19, Ex-Ger'n Cargo	8749	1400	1400	0	0	0
134	Sol Navis / June 25-19, Am-Cargo	11075	2400	2400	0	0	0
135	South Bend / May 5-19, Am-Cargo	17716	2300	2300	0	0	0
136	Suwanee / Apr. 11-19, Ex-Ger'n Cargo	6000	2000	2000	0	0	0
137	Texan / Jan. 18-19, Am-Cargo	19000	2200	2200	0	0	0
138	Tiger / Mar. 7-19, Am-Cargo	10000	2600	2600	0	0	0
139	Troy / Feb. 27-19, Am-Cargo	37336	5900	5900	0	0	0
140	Virginian / Feb. 1-19, Am-Cargo	12600	4000	4300	0	0	0
141	Yale / June 15-19, Am-Cargo	10000			0	0	0
	Total	884008	140235	146035	0	0	0

German Ships Used for Returning

No.	Name of Ship / Date Placed in Commission or Attached to Force / Type of Vessel	Displacement	Original Troop Carrying Capacity Including Officers	Maximum Troop Carrying Capacity Including Officers	Voyages Made Prior to Signing of Armistice		
					Number of Turn-arounds Made	Total Number of All Passengers Carried to Europe	Total Number of Passengers Returned from Europe
142	Cap Finsterre / Apr. 11-19, German Pass'r	23000	3800	3800	0	0	0
143	Graf Waldersee / Mar. 28-19, German Pass'r	13193	4300	4300	0	0	0
144	Imperator / May 5-19, German Pass'r	60000	8900	9800	0	0	0
145	K. A. Victoria / Apr. 27-19, German Pass'r	30400	5500	5500	0	0	0
146	Mobile / Mar. 26-19, German Pass'r	27000	4800	5200	0	0	0
147	Patricia / Apr. 25-19, German Pass'r	12500	2900	2900	0	0	0
148	Pretoria / Aug. 24-19, German Pass'r	14100	3000	3000	0	0	0
149	P. F. Wilhelm / Mar. 30-19, German Pass'r	26050	3600	3600	0	0	0
150	Zeppelin / Mar. 29-19, German Pass'r	12450	4300	4300	0	0	0
	Total	220669	41100	42400	0	0	0
	GRAND TOTAL	2341038	332085	383935	306	870324	5051

Table - Cruiser and Transport Force

CRUISERS ENGAGED IN TRANSPORTING TROOPS TO AND FROM FRANCE BETWEEN WERE OPERATED UNDER THE COMMAND OF THE COMMANDER OF TRANSPORT FORCE

during the War and continued in service after the Armistice was signed and were troops back from France

Number of Turn-arounds Made	Total Number of All Passengers Carried to Europe	Total Number of All Passengers Returned from Europe	Total Number of Sick and Wounded Returned from Europe	Total Number of All Passengers Carried to and from Europe	Final Disposition — Date of Arrival in U. S. on Last Voyage as a Transport — Date Placed out of Commission or Transferred from Force
7	182	22080	5018	47432	Shipping Board Sept. 5–19—Sept. 5–19
9	1782	41179	4425	78249	Army Transport Service Aug. 18–19—Aug. 27–19
8	42	46823	4668	86801	Army Transport Service Sept. 15–19—Sept. 26–19
8	13	22065	4150	38705	Army Transport Service Sept. 15–19—Sept. 24–19
5	41	10113	21	17821	United Fruit Co. Aug. 17–19—Aug. 19–19
0	0	0	0	21628	Torpedoed and sunk, July 1, 1918
8	1	8949	3868	20332	Shipping Board Sept. 5–19—Sept. 6–19
8	11	27762	4435	40443	Inter. Mercantile Marine Sept. 4–19—Sept. 4–19
9	' 351	34142	5085	83350	
8	2308	22852	5522	54085	Army Transport Service Aug. 8–19—Aug. 15–19
0	0	0	0	1438	June 4–18—Sept. 7–19
6	624	14140	2808	24619	Inter. Mercantile Marine Aug. 28–19—Aug. 11–19
6	822	8606	4284	25892	Aug. 25–19—Sept. 12–19
7	138	20582	1546	41658	Shipping Board Aug. 23–19—Aug. 25–19
6	0	11339	1296	17622	Dutch Government Aug. 19–19—Aug. 19–19
8	22	23598	2554	37822	Shipping Board Sept. 10–19—Sept. 13–19
0	0	0	8	8975	United Fruit Co. Sept. 3–18—Oct. 29–18

Number of Turn-arounds Made	Total Number of All Passengers Carried to Europe	Total Number of All Passengers Returned from Europe	Total Number of Sick and Wounded Returned from Europe	Total Number of All Passengers Carried to and from Europe	Final Disposition Date of Arrival in U. S. on Last Voyage as a Transport Date Placed out of Commission or Transferred from Force
9	1517	93746	10913	192753	Shipping Board Sept. 8-19—Sept. 9-19
7	166	14823	1538	24250	Inter. Mercantile Marine Aug. 20-19—Aug. 20-19
7	7	16978	2237	34937	Army Transport Service Aug. 23-19—Sept. 2-19
7	2	12143	2371	21901	Mallory S. S. Co. Aug. 29-19—Aug. 30-19
9	232	39501	6186	54230	Atlantic Transport Co. Aug. 25-19—Aug. 29-19
8	127	19201	987	41824	July 27-19
8	237	23321	853	36895	Matson Navigation Co. Aug. 20-19—Aug. 21-19
8	3	25217	8184	36273	Matson Navigation Co. Aug. 17-19—Aug. 18-19
8	30	20871	510	39463	Army Transport Service Sept. 19-19—Sept. 27-19
8	487	34813	2707	54337	Atlantic Transport Co. Aug. 9-19—Aug. 18-19
8	125	42500	4015	76402	Army Transport Service Sept. 11-19—Sept. 29-19
4	0	8117	5895	28866	Army Transport Service Aug. 12-19—Aug. 21-19
9	16	31705	2933	47449	Army Transport Service Aug. 30-19—Sept. 4-19
8	0	14000	4597	24027	Shipping Board Aug. 30-19—Aug. 30-19
7	509	14634	2956	24330	Inter. Mercantile Marine Aug. 29-19—Aug. 29-19
9	1715	20693	1382	45141	
6	46	15392	1880	30087	Army Transport Service Aug. 23-19—Sept. 2-19
8	130	37025	3301	77129	Army Transport Service Sept. 22-19—Oct. 6-19
0	0	0	9	20143	Torpedoed and sunk, May 31, 1918
8	2015	24859	5251	48296	Army Transport Service Sept. 10-19—Sept. 16-19
7	5	20972	4465	39329	Dutch Government Aug. 4-19—Aug. 4-19

Number of Turn-arounds made	Total Number of All Passengers carried to Europe	Total Number of All Passengers Returned from Europe	Total Number of Sick and Wounded Returned from Europe	Total Number of All Passengers Carried to and from Europe	Final Disposition — Date of Arrival in U. S. on Last Voyage as a Transport — Date Placed out of Commission or Transferred from Force
10	177	34702	5307	55169	Army Transport Service Sept. 2-19—Sept. 10-19
8	3	10689	2250	12404	Oceanic S. S. Co. Sept. 1-19—Sept. 1-19
7	1029	15537	2676	34911	Shipping Board Aug. 27-19—Aug. 29-19
1	0	1664	226	17370	Stranded on rocks at St. Nazaire, Dec. 28, 1918
8	1187	22025	2253	37580	Army Transport Service Sept. 28-19—Oct. 13-19
7	3	11577	2610	22723	Matson Navigation Co. Aug. 6-19—Aug. 6-19
7	3170	15737	3549	27344	Dutch Government July 31-19—July 31-19
304	19275	956672	141779	1850435	

USED FOR RETURNING TROOPS

Number of Turn-arounds Made	Total Number of All Passengers Carried to Europe	Total Number of All Passengers Returned from Europe	Total Number of Sick and Wounded Returned from Europe	Total Number of All Passengers Carried to and from Europe	Final Disposition — Date of Arrival in U. S. on Last Voyage as a Transport — Date Placed out of Commission or Transferred from Force
5	0	7704	34	7704	Returned to Fleet June 29-19—July 2-19
4	1	4861	30	4862	Returned to Fleet June 22-19—June 23-19
6	2	9659	83	9661	Returned to Fleet July 12-19—July 14-19
5	0	5869	58	5869	Returned to Fleet June 28-19—July 1-19
6	0	11913	42	11913	Returned to Fleet July 5-19—July 8-19

Number of Turn-arounds made	Total Number of All Passengers carried to Europe	Total Number of All Passengers Returned from Europe	Total Number of Sick and Wounded Returned from Europe	Total Number of All Passengers Carried to and from Europe	Final Disposition Date of Arrival in U. S. on Last Voyage as a Transport Date Placed out of Commission or Transferred from Force
Voyages Made from Signing of Armistice to Oct. 1, 1919					
5	0	7486	83	7486	Returned to Fleet June 27–19—July 1–19
4	0	4714	29	4714	Returned to Fleet June 30–19—June 30–19
2	0	1052	22	1052	Returned to Fleet Apr. 26–19—July 2–19
3	0	3955	12	3955	Returned to Fleet July 21–19—July 29–19
4	0	3278	14	3278	Returned to Fleet July 26–19—July 28–19
6	1	8800	29	8801	Returned to Fleet June 30–19—July 3–19
4	10	4530	47	4540	Returned to Fleet June 21–19—June 22–19
4	2	4900	14	4902	Returned to Fleet June 22–19—June 24–19
4	0	4675	29	4675	Returned to Fleet June 7–19—June 9–19
6	0	8962	15	8962	Returned to Fleet July 1–19—July 3–19
1	0	778	8	778	Returned to Fleet Mar. 13–19—Mar. 15–19
6	0	10136	33	10136	Returned to Fleet July 13–19—July 15–19
5	0	5303	26	5303	Returned to Fleet July 4–19—July 6–19
1	0	317	0	317	Returned to Fleet Mar. 4–19—Mar. 4–19
6	1	9397	14	9398	Returned to Fleet July 4–19—July 6–19
4	1	4501	11	4502	Returned to Fleet July 26–19—July 28–19
2	0	3463	0	3463	Returned to Fleet July 19–19—July 20–29
6	0	8437	22	8437	Returned to Fleet July 13–19—July 14–19
4	0	4795	18	4795	Returned to Fleet June 20–19—June 22–19
5	0	5784	18	5784	Returned to Fleet July 5–19—July 7–19
108	18	145249	681	145287	

Number of Turn-arounds Made	Total Number of All Passengers Carried to Europe	Total Number of All Passengers Returned from Europe	Total Number of Sick and Wounded Returned from Europe	Total Number of All Passengers Carried to and from Europe	Final Disposition Date of Arrival in U. S. on Last Voyage as a Transport Date Placed out of Commission or Transferred from Force
4	0	8643	35	8643	Am-Hawaiian Co. July 16–19—July 16–19
3	0	6417	45	6417	Shipping Board Sept. 3–19—Sept. 4–19
2	0	6112	40	6112	Panama R. R. Co. July 7–19—July 15–19
5	0	4700	40	5700	Shipping Board Sept. 11–19—Sept. 13–19
4	0	7794	28	7794	Am-Hawaiian Co. Sept. 2–19—Sept. 2–19
4	0	11760	120	11760	Shipping Board Sept. 23–19—Sept. 24–19
3	0	4759	25	4759	Shipping Board July 21–19—Aug. 1–19
5	0	4717	24	4717	Army Trans. Service Aug. 22–19—Aug. 26–19
2	0	3731	52	3731	Shipping Board Sept. 4–19—Sept. 8–19
4	0	4828	32	4828	S. Pacific R. R. Co. Aug. 26–19—Aug. 26–19
3	0	4153	27	4153	S. Pacific R. R. Co. July 10–19—July 12–19
3	1	5726	19	5727	Shipping Board July 5–19—July 14–19
3	0	1192	649	1192	U. S. Navy Mar. 13–19—Mar. 13–19
5	5	8812	37	8817	Am-Hawaiian Co. July 20–19—July 20–19
1	0	985	3	985	Army Trans. Service Sept. 4–19—Sept. 12–19
2	0	2710	4	2710	Ward Line Aug. 23–19—Aug. 23–19
2	0	2981	5	2981	Ward Line Aug. 24–19—Aug. 25–19
2	0	3296	46	3296	Shipping Board July 31–19—Aug. 5–19
2	0	1886	0	1886	Nafia S. S. Co. Sept. 14–19—Sept. 27–19

Number of Turn-arounds Made	Total Number of All Passengers Carried to Europe	Total Number of All Passengers Returned from Europe	Total Number of Sick and Wounded Returned from Europe	Total Number of All Passengers Carried to and from Europe	Final Disposition Date of Arrival in U. S. on Last Voyage as a Transport Date Placed out of Commission or Transferred from Force
4	0	7209	19	7209	Am-Hawaiian Co. July 15-19—July 17-19
4	2	4981	5	4983	Shipping Board Sept. 4-19—Sept. 5-19
4	0	4238	20	4238	Panama R. R. Co. July 8-19—Aug. 27-19
2	19	2063	13	2082	Panama R. R. Co. July 3-19—July 15-19
3	0	4166	6	4166	S. Pacific R. R. Co. July 13-19—July 15-19
6	0	9876	32	9876	Am-Hawaiian Co. Aug. 29-19—Aug. 30-19
5	0	8895	23	8895	Am-Hawaiian Co. Aug. 30-19—Sept. 2-19
4	0	5624	276	5624	Shipping Board Sept. 4-19—Sept. 5-19
5	0	9658	9	9658	Shipping Board Sept. 4-19—Sept. 4-19
5	0	9372	13	9372	
3	0	6812	28	6812	Luckenbach Co. July 3-19—July 28-19
2	0	4695	12	4695	Luckenbach Co. July 29-19—July 31-19
4	0	10579	39	10579	Luckenbach Co. Aug. 4-19—Aug. 4-19
1	0	1	0	1	Luckenbach Co.
3	0	4833	10	4833	Luckenbach Co. Sept. 15-19—Sept. 13-19
5	0	12525	300	12525	Luckenbach Co. July 11-19—July 17-19
2	0	3243	8	3243	Army Trans. Service Sept. 2-19—Sept. 12-19
4	143	1946	1977	2089	U. S. Navy May 25-19—May 25-19
5	0	12386	37	12386	Am-Hawaiian July 23-19—July 23-19

Number of Turn-arounds Made	Total Number of All Passengers Carried to Europe	Total Number of All Passengers Returned from Europe	Total Number of Sick and Wounded Returned from Europe	Total Number of All Passengers Carried to and from Europe	Final Disposition Date of Arrival in U. S. on Last Voyage as a Transport Date Placed out of Commission or Transferred from Force
	Voyages Made from Signing of Armistice to Oct. 1, 1919				
4	0	8038	164	8038	Am-Hawaiian Aug. 3–19—Aug. 4–19
4	0	7587	15	7587	Shipping Board Sept. 10–19—Sept. 15–19
5	0	23619	557	23619	Shipping Board Aug. 23–19—Aug. 25–19
6	0	8383	42	8383	Am-Hawaiian Co. Sept. 16–19—Sept. 16–19
4	6	3446	79	3452	Shipping Board Aug. 28–19—Aug. 28–19
6	0	11393	26	11393	Am-Hawaiian Co. Aug. 29–19—Aug. 29–19
2	0	2736	4	2736	Shipping Board July 14–19—July 16–19
3	0	4659	11	4659	Standard Trans. Co. Aug. 30–19—Sept. 2–19
2	0	4142	6	4142	Shipping Board Sept. 26–19—Sept. 23–19
4	0	5876	15	5876	Shipping Board Sept. 23–19—Sept. 24–19
4	0	5507	19	5507	S. Pacific R. R. Co. Aug. 1–19—Aug. 1–19
4	1	5960	39	5961	Grace S. S. Co. July 7–19—July 14–19
4	0	6310	6	6310	Grace S. S. Co. July 23–19—July 24–19
4	0	6126	67	6126	Nafia S. S. Co. Sept. 7–19—Sept. 26–19
4	0	6863	11	6863	Atlantic & Pacific Co. Aug. 3–19—Aug. 3–19
2	0	1707	3	1707	Cunard S. S. Co. July 23–19—Aug. 20–19
2	0	2312	6	2312	Shipping Board Sept. 19–19—Sept. 26–19
1	0	395	26	395	Army Trans. Service Aug. 19–19—Sept. 9–19
3	0	3756	21	3756	Shipping Board Aug. 30–19—Oct. 14–19

Voyages Made from Signing of Armistice to Oct. 1, 1919			Total Number of Sick and Wounded Returned from Europe	Total Number of All Passengers Carried to and from Europe	Final Disposition
Number of Turn-arounds Made	Total Number of All Passengers Carried to Europe	Total Number of All Passengers Returned from Europe			Date of Arrival in U. S. on Last Voyage as a Transport; Date Placed out of Commission or Transferred from Force
4	0	7491	14	7491	Atlantic & Pacific Co. July 9-19—July 14-19
4	2	7447	172	7449	Grace S. S. Co. Aug. 4-19—Aug. 4-19
4	0	6302	29	6302	Grace S. S. Co. Sept. 23-19—Sept. 24-19
8	0	14264	4518	14264	Grace S. S. Co. Sept. 4-19—Sept. 8-19
3	0	5625	15	5625	Shipping Board July 16-19—July 16-19
2	0	2820	4	2820	Shipping Board July 16-19—July 18-19
2	0	3264	8	3264	Shipping Board Sept. 26-19—Sept. 29-19
3	0	4875	110	4875	Army Trans. Service Aug. 23-19—Sept. 3-19
3	0	4801	15	4801	Shipping Board Sept. 3-19—Sept. 3-19
4	3	8668	7	8671	Standard Trans. Co. Aug. 5-19—Aug. 7-19
3	0	7739	55	7739	Standard Trans. Co. July 29-19—July 29-19
3	4	14039	45	14043	Standard Trans. Co. Aug. 20-19—Aug. 21-19
4	0	16631	279	16631	Am-Hawaiian Aug. 3-19—Aug. 4-19
1	0	901	0	901	U. S. Navy June 20-19—July 20-19
246	186	441986	10452	442172	
3½	1121	9718	58	10839	Shipping Board Aug. 19-19—Sept. 29-19
2½	0	7728	21	7728	Shipping Board Aug. 30-19—Sept. 27-19
3½	161	28030	147	28191	Shipping Board Aug. 10-19—Sept. 19-19
4½	31	22674	460	22705	Shipping Board Aug. 22-19
4½	12	21073	22	21085	Shipping Board Sept. 3-19—Sept. 30-19
3½	0	8572	11	8572	Shipping Board Aug. 16-19—Sept. 12-19
3½	1083	10364	40	11447	Shipping Board Aug. 31-19—Sept. 29-19
4½	1	14161	21	14162	Shipping Board Aug. 23-19—Sept.-
4½	0	15800	28	15800	Shipping Board Sept. 5-19—Oct. 1-19
34½	2409	138120	808	140529	
692½	21888	1082027	153720	2578423	

Record of Leading Ships

1	Leviathan July 25–17, Ex-German	69000	9000	12000	10	96804	686
2	America Aug. 6–17, Ex-German	41500	4000	7000	9	37768	168
3	George Washington Sept. 6–17, Ex-German	39435	5600	6500	9	48373	484
4	Agamemnon Aug. 21–17, Ex-German	30000	3400	5800	10	36097	214
5	Pres. Grant Aug. 2–17, Ex-German	33000	4800	5900	8	39974	0
6	Mount Vernon July 28–17, Ex-German	32130	3100	5800	9	33692	86
7	Siboney Apr. 8–18, Am-Passenger	11250	3100	4000	7	20299	11
8	Mongolia May 8–18, Am-Passenger	26695	3700	4700	5	19013	24
9	Manchuria Apr. 25–18, Am-Passenger	26500	3500	4800	4	14491	16
10	Great Northern Nov. 1–17, Am-Passenger	14000	2800	3300	10	28248	677
	Total	323510	43000	59800	81	374679	2366

Table F

Number of Turn-arounds Made	Voyages Made from Signing of Armistice to Oct. 1, 1919		Total Number of Sick and Wounded Returned from Europe	Total Number of All Passengers Carried to and from Europe	Final Disposition Date of Arrival in U. S. on Last Voyage as a Transport Date Placed out of Commission or Transferred from Force
	Total Number of All Passengers Carried to Europe	Total Number of All Passengers Returned from Europe			
9	1517	93746	10913	192753	U. S. Shipping Board Sept. 8–19—Sept. 9–19
8	42	46823	4668	86801	Army Trans. Service Sept. 15–19—Sept. 26–19
9	351	34142	5085	83350	
9	1782	41179	4425	78249	Army Trans. Service Aug. 18–19—Aug. 27–19
8	130	37025	3301	77129	Army Trans. Service Sept. 22–19—Oct. 6–19
8	125	42500	4015	76402	Army Trans. Service Sept. 11–19—Sept. 29–19
10	177	34702	5307	55169	Army Trans. Service Sept. 2–19—Sept. 10–19
8	487	34813	2707	54337	Atlantic Trans. Co. Aug. 9–19—Aug. 18–19
9	232	39501	6186	54230	Atlantic Trans. Co. Aug. 25–19—Aug. 29–19
8	2308	22852	5522	54085	Army Trans. Service Aug. 8–19—Aug. 15–19
86	7151	427283	52129	812505	

Table G

1918

1918	ARMY 151649						
Month	Mobile	Litter	G. U.	Insane	T. B.	Contag.	Dead
January	85	7	3	6	22		7
February	29	18		7	82	1	
March	78	2	4	12	19	10	4
April	59	27	16	16	66	42	26
May	148	39	7	66	89	24	7
June	95	69	5	29	50	6	4
July	349	204	23	137	140	39	5
August	505	180	49	213	162	56	12
September	1667	537	40	336	143	14	3
October	2701	1005	23	593	229	333	322
November	5718	877	67	175	224	152	39
December	14786	1335	73	500	261	55	15
Total	26220	4300	310	2091	1487	732	444

1918	NAVY 4395						
January	16				2		
February	4			1	1		3
March	15	1					3
April	7	9	13		3	9	4
May	16	2	5	2	4	2	
June	17		15	4	4	2	3
July	36	15	16	9	7	9	2
August	59	32	30	15	20	8	8
September	24	14	11	11	3	4	40
October	58	14	21	6	15	27	30
November	140	15	25	3	6	3	4
December	219	6	58	7	13	7	7
Total	611	108	194	58	78	71	102

1919	Army 151649						
Month	Mobile	Litter	G. U.	Insane	T. B.	Contag.	Dead
January	15520	2054	66	382	303	158	9
February	13019	1020	70	421	292	470	31
March	19203	2331	90	796	533	750	38
April	15163	1902	60	645	754	389	12
May	14961	1019	78	1226	600	231	18
June	10921	910	207	506	221	91	12
July	3361	635	322	383	207	46	11
August	2343	217	108	107	94	65	1
September	364	206	23	46	36	6	2
Total	94855	10294	1024	4512	3040	2206	134
GRAND TOTAL	121075	14594	1334	6603	4527	2938	578

1919	Navy 4395						
January	773	86	67	9	17	4	
February	190	40	109	3	6	27	6
March	188	47	82	5	12	24	4
April	138	53	50	4	11	38	4
May	169	37	72	5	4	15	2
June	132	30	37	6	12	4	1
July	76	16	64	3	3	3	1
August	76	26	117	2	3	8	5
September	112	23	98	10	4		
Total	1854	358	696	47	72	123	23
GRAND TOTAL	2465	466	890	105	150	194	125

1918

1918	Marines 3626							
Month	Mobile	Litter	G. U.	Insane	T. B.	Contag.	Dead	Total Army, Navy and Marines
January								148
February					2			148
March	3						1	152
April				2				299
May	1				2			414
June		2		1	2			308
July	9	7		1	3			1011
August	48	13			3		1	1414
September	124	69	1	1	3			3045
October	136	59	4	5	3	2	53	5639
November	182	50			9		6	7695
December	337	51	15	6	3		9	17763
Total	840	251	20	16	30	2	70	38036

1919	MARINES 3626							
Month	Mobile	Litter	G. U.	Insane	T. B.	Contag.	Dead	Total Army, Navy and Marines
January	139	14	3	5	2			19611
February	235	8	2	2	2			15933
March	560	51	3	13	9	8		24748
April	381	42		14	6	1		19646
May	312	18		17	9	25		18818
June	239	9	3	10	4			13355
July	50	2		3	2			5188
August	142	8	15	7	2	4		3350
September	11	4						945
Total	2069	156	26	71	36	38		121634
GRAND TOTAL	2909	407	46	87	66	41	70	159670

Memorandum of Admiral von Holtzendorff, Chief of the German Admiralty

To B 35840 I

Berlin, Dec. 22, 1916.

(Strictly secret)

I have the honor to transmit to Your Excellency in the annex a note on the necessity of a speedy commencement of the unrestricted U-boat war.

Based on the detailed explanations of the annex, I may beg Your Excellency to consider the following ideas, and I hope to gain a complete agreement in our opinions that it is absolutely necessary to intensify to the utmost possibility our measures against England's sea traffic in order to take advantage of the favorable situation and to secure for us a speedy victory.

The war requires a decision before Autumn, 1917, if it is not to end in a general exhaustion of all parties, which would be fatal for us too. Among our adversaries, the economical conditions of Italy and France have been so seriously shaken that they can only be maintained by the energy and strength of England. If we succeed in overcoming England the war will be decided at once in our favor. But the resource of England is her tonnage, which supplies the islands of Great Britain with the necessities for life and the war industry and at the same time secures her solvency abroad.

SAW GREAT SHORTAGE OF SHIPS

The present state of the tonnage question is in short as follows:

The freight for a great number of important goods has risen enormously, in certain places to tenfold amount and more. We also know for certain from numerous other proofs that the lack of tonnage is universal.

The English tonnage at present still existing may be reckoned to be about 20 million gross register tons. At least 8.6 million tons of these are requisitioned for military purposes and one-half million tons is employed in coastal traffic; approximately one million tons is under repair or temporarily out of use; about two million tons are used in the interest of the Allies; so that, at the highest, eight million tons of British tonnage are at the disposal of England's supplies.

A perusal of the statistics of the sea traffic in English harbors would return even a lower figure. Thus in the months of July-September, 1916, there were only 6 ¾ million gross register tons of British tonnage available for England. Apart from this, the other tonnage bound for England may be calculated at 900,000 tons of enemy tonnage, none English, and quite three million tons of neutral tonnage. All in all, England is therefore supplied by only just 10 ¾ million gross register tons.

Besides the fact that, based on the achievements hitherto performed in the struggle against the tonnage, it seems to be very promising for us to proceed on the way once taken. The unusually bad result of this year's world harvest in cereals and cattle food has given us a unique opportunity, which cannot be neglected by any one with a sense of responsibility. Already after February the United States and Canada will probably be unable to provide England with corn, therefore England must procure her supply from over long distances, Argentina, and as Argentina can supply only a little on account of its bad harvest, she will be compelled to import from India and chiefly from Australia.

FORCE PEACE WITHIN FIVE MONTHS

Under such favorable conditions an energetic powerful blow against the English tonnage promises to have an absolutely certain success. I do not hesitate to declare that, under the prevailing conditions, we may force England into peace within five months through the unrestricted U-boat war. However, this can only be achieved by the unrestricted U-boat war, not by the U-boat cruising as practiced at present, and not even if all armed vessels were free to be sunk.

Based on the formerly mentioned monthly rate of destruction of 600,000 tons of tonnage by the unrestricted U-boat war, and on the expectation that by it at least two-fifths of the neutral traffic will be frightened to undertake the voyage to England, it may be reckoned that the English sea traffic after five months will be reduced by about 39 per cent of the traffic.

England would not be able to bear this, neither in view of the conditions after the war nor as regards the possibility of continuing the war. She is now already facing a scarcity of food, which forces her to try measures of economy which we, as a blockaded country, had to adopt during the war. The conditions for such an organization are totally different in England and comparatively much more unfavorable than with ourselves. There are lacking authorities as well as the sense of the people to submit to such force.

Also from another cause the general reduction of the bread ration for the whole population cannot now be enforced in England. This measure was possible in Germany at a time when temporarily other foodstuffs could make good the sudden reduction of the bread ration.

TOLD OF LOW BRITISH SUPPLIES

This opportunity has been allowed to pass and cannot possibly be brought back. But the maintenance of the war industry, and at the same time that of the food supply, cannot be kept up with about three-fifths of the sea traffic, without universal severe rationing of the consumption of cereals. The argument that England might have sufficient grain and raw materials in the country in order to overcome the danger until the next harvest is refuted exhaustively in the annex.

In addition, the unrestricted U-boat war with the subsequent cessation of supply by Denmark and Holland would mean for England at once the scarcity of fat, as one-third of the whole British import of butter originates from Denmark, and the entire supply of margarine comes from Holland. Furthermore, it would mean the severity of the lack of raw materials and wood by endangering the supply of these products from Scandinavia and at the same time increasing the attenuation of the Spanish supply of metal.

Finally we shall have the long wished for opportunity to deal with the neutral supply of ammunition and thus relieve somewhat the army. (These ammunition supplies came chiefly from America.)

In the face of such facts the U-boat war, as practiced hitherto, would even after general permission to sink all armed vessels result in five months' time in the diminution of all the tonnage bound for England by only 5,400,000 tons — viz., about 18 per cent, of the present monthly sea traffic, therefore less than one-half what could be obtained by the unrestricted U-boat war.

PANIC ESSENTIAL TO SUCCESS

In addition, the lack of psychological effects of panic and terror is to be considered. I regard these effects, expected only by the unrestricted U-boat war, as an essential preconception of success. The experiences gained at the beginning of the U-boat war after the Spring of 1915, when the English still believed its bitter seriousness, and even in the short U-boat war of March and April, 1916, proved how weighty these effects are.

Moreover, a preliminary condition is that the beginning and the declaration of the unrestricted U-boat war must follow so quickly one upon the other that there is no time for negotiations, especially between England and the neutrals. The wholesome terror will exercise in this case upon enemy and neutral alike.

The declaration of the unrestricted U-boat war will place before the Government of the United States of North America afresh the question whether or not she will take the consequences of her hitherto adopted attitude toward the use of U-boats. I am quite of opinion that the war against America is so serious an affair that all must be done to avert it. However, the dread of a break must not, in my opinion, go so far as to make us shrink in the decisive moment from the use of the weapon which will bring us victory.

At any rate it will be expedient to consider what influence the entrance of America into the war on the side of our adversaries would have upon the trend of the war.

As regards tonnage, this influence would be very negligible. It is not to be expected that more than a small fraction of the tonnage of the Central Powers lying in America and many other neutral harbors could then be enlisted for the traffic to England.

ALREADY ORDERED SHIPS DAMAGED

For the far greatest part of this shipping can be damaged in such a way that it cannot sail in the decisive time of the first months. Preparations to this effect have been made. There would also be no crews to be found for them. Just as little decisive effect can be ascribed to any considerable extent to American troops, which, in the first place, cannot be brought over, through lack of tonnage.

There remains only the question, what attitude would America take in the face of a conclusion of peace into which England would be coerced! It is not to be supposed that she would then decide to continue the war, as she would have no means at her disposal to take any decisive action against us, while her sea traffic will be liable to be damaged by us. On the contrary, it is to be expected that she will participate in the English conclusion of peace in order to obtain as quickly as possible again sound economic conditions.

I therefore draw the conclusion that an unrestricted U-boat war, which must be recommended as early as possible in order to bring about peace before the world's harvest of Summer, 1917, that is, before August 1st, should even take the consequences of a break with America, because we have no other alternative. A quickly launched, unrestricted U-boat war is therefore the only correct means to end the war victoriously, in spite of the risk of a break with America. It is also the only way to this goal.

In order to obtain in due time the necessary effect, the unrestricted U-boat war must commence at the latest on February 1st. I beg Your Excellency to inform me whether the military situation on the Continent, especially in the face of the still remaining neutrals, will permit of this date. I require a period of three weeks in order to make the necessary preparations.

V. HOLTZENDORFF.

Officer Personnel of United States Cruiser and Naval Transport Service

FROM MOBILIZATION IN 1917 TO THE ARMISTICE, NOVEMBER, 1918

CRUISER AND TRANSPORT FORCE

FORCE COMMANDER
VICE ADMIRAL ALBERT GLEAVES, U. S. N.
FLAGSHIP
U. S. S. SEATTLE
CHIEF OF STAFF
CAPTAIN DE W. BLAMER, U. S. N.

FORCE TRANSPORT OFFICE

Captain A. H. Robertson, USN
Captain C. B. Morgan, USN
Commander Robert Henderson, USN
Commander C. C. Soule, USN
Commander W. S. Giles, USN
Commander E. Armstrong, USN

FORCE ENGINEER OFFICE

Lieut.-Comdr. F. M. Perkins, USN
Lieut.-Comdr. E. D. Almy, USN
Lieutenant (T) S. L. Almon, USN
Lieutenant C. E. Milbury, USNRF
Boatswain (T) W. E. McCabe, USN

FORCE SUPPLY OFFICE

Comdr. (SC) Ray Spear, USN
Ensign (SC) O. Tagland, USN
Ensign (SC) J. D. Gagan, USN

FORCE MEDICAL OFFICE

Comdr. (MC) A. L. Clifton, USN
Comdr. (MC) C. N. Fiske, USN
Comdr. (MC) J. J. Snyder, USN
Lieut. (DC) J. V. McAlpin, USN

FLAG SECRETARIES

Lieut.-Comdr. A. L. Bristol,
USN Commander C. C. Gill, USN

FLAG LIEUTENANTS

Lieut.-Comdr. T. A. Symington, USN
Lieutenant J. H. Lawson, USN

TORPEDO AND GUNNERY OFFICE
(Also Personnel)

Lieut.-Comdr. F. H. Roberts, USN
Lieutenant E. N. Fisher, USN

FORCE RADIO

Lieutenant C. N. Ingraham, USN
Lieutenant R. S. H. Venable, USN
Lieutenant H. L. Leeb, USNRF

FORCE MARINE OFFICE

Captain R. H. Tebbs, Jr., USMC

COMMUNICATIONS

Lieutenant J. P. Brown, USN
Lieutenant J. S. Watters,
USN Lieutenant Wm. H. Long, USNRF
Lieutenant De C. Fales, USNRF
Ensign O. B. Jennings, USNRF
Ensign M. P. Sherwood, USNRF
Ensign H. L. Willoughby, USNRF
Ensign (T) E. H. Wardwell, USNRF

NEWPORT NEWS DIVISION OF CRUISER AND TRANSPORT FORCE

Commander: Rear Admiral H. P. Jones, U. S. N.
Flagship: U. S. S. *SIALIA*
Staff:
Captain J. F. Hines, IT. S. N. ....Chief of Staff
Lieut. Commander K. H. Donavin, U. S. N............ Aid
Lieutenant S. W. King, U. S. N................................Aid
Lieutenant T. S. King, 2nd U. S. N..........................Aid

Additional Officers in Order of Reporting for Duty

Ensign S. P. Sears, U.S.N.R.F..Aid
Lieutenant (j.g.) H. R. Wakeman, U.S.N.R.F..........................Aid
Dental Surgeon C. E. Detmer, U.S.N.R.F..................................Aid
Dental Surgeon W. J. Davidson, U. S. N..................................Aid
Paymaster B. Mayer, U. S. N.................Division Supply Officer
Ensign (T) H. M. Leisure, U. S. N..Aid
Lieutenant G. H. Jett, U.S.N.R.F..........Division Engineer Officer
Ensign (T) H. B. Leland, U. S. N..Aid
Ensign I. B. Levi, U.S.N.R.F...Aid
Ensign W. J. Murray, U.S.N.R.F..Aid
Pharmacist W. M. Benton, U. S. N...Aid
Ensign P. Seay, U.S.N.R.F..Aid

CRUISER FORCE: SQUADRON 2

Commander: Rear Admiral Marbury Johnston, U. S. N.
Flagship: U. S. S. *ISIS*
Staff:
Lieutenant Commander R. S. Galloway, U. S. N.....Aid
Lieutenant (j.g.) C. A. MacDonald, N. N. V.
Ensign H. W. Browne, N. N. V.

Additional Officers in Order of Reporting for Duty

Lieutenant (j.g.) H. R. Leonard, U.S.N.R.F.
Lieutenant Commander J. L. Duffy, U.S.N.R.F.
Lieutenant Commander E. C. Jones, U.S.N.R.F.
Lieutenant Commander L. E. Congdon, U.S.N.R.F.
Lieutenant Commander R. B. Powers, U.S.N.R.F.
Lieutenant Commander M. J. Flannagan, U.S.N.R.F.
Lieutenant Commander W. A. Hogan, U.S.N.R.F.
Lieutenant Commander R. Mod. Moser, U.S.N.R.F.
Lieutenant Commander L F. Shurtleff, U.S.N.R.F.

U. S. S. Aeolus

(Transport)

C. S. Kempff, Commander, USN
M. W. Hutchinson, Jr., Ensign, USN
F. F. Ingram Ensign (T), USN
H. L. Smith, P. A. Surgeon, USN
W. A. Fort, Asst. Surgeon, USNRF
F. P. James, Asst. Surgeon, USNRF
H. R. Snyder, P. A, Paymaster, USN
C, M. Austin, Lieutenant, USN
J. D. Hashagen, Jr., Lieutenant, USNRF
A. J. Shrader, Lieutenant (jg), USNRF
J. C. McDermott, Ensign (T), USN
W. R. Gardner, Ensign (T), USN
O. S. Dynes, Ensign, USNRF
A. B. Torrey, Ensign, USNRF
E. H. Tricou, Paymaster, USN
L. S. Hill, Asst. Paymaster, USNRF
P. Mullen, Chief Bosn., USN
B. C. Phillips, Boatswain (T), USN
J. H. Dwyer, Machinist, USNRF
H. F. Helmken, Machinist, USNRF
G. H. Wheeler, Carpenter, USN
B. H. White, Pay Clerk, USN

Relief Officers in Order of Reporting on Board

H. V. McCabe, Lieutenant (jg), USN
G. E. Wiebe, Lieutenant, USNRF
C. H. Zearfoss, Lieutenant (jg), USNRF
R. Snyder, Ensign (T), USN
W. J. Wheatley, Ensign, USNRF
M.W. Boykin, Asst. Paymaster, USNRF
R, J. White, Asst. Paymaster, USN
L. C. Fuller, Act. Pay Clerk, USN
N. C. Lovegrove, Ensign, USNRF
T. M. Possatt, Ensign, USNRF
C. E. Evans, Gunner (T), USN
H. G. Mecklenberg, Machinist (T) USN
J. A. Sherman, Carpenter (T), USN
W. E. G. Bartle, Pharmacist, USN
J. Metayer, Lieut. Comdr., USN (Ret.)
S. J. Skou, Ensign (T), USN
L. McCormick, Ensign (T), USN
C. E. Rockwell, Ensign, USNRF
A. A. Walker, Ensign, USNRF
L. E. Walker, Ensign, USNRF
B. Weston, Ensign, USNRF
W. J. Wheatley, Ensign, USNRF

B. E. Jolidan, Ensign, USNRF
L. G. Fuller, Asst. Paymaster (T), USN
J. W. Decker, Act. Chaplain, USN
H. B. Grounds, Bosn. (T), USN
N. S. Winskill, Lieutenant, USNRF
H. W. Olds, Ensign, USN
C. E. Olsen, Ensign, USN
J. L. Bigelow, Ensign, USNRF
D. A. Frieman, Ensign, USNRF
W. C. Richer, Ensign, USNRF
E. W. Reynolds, Ensign, USNRF
F. E. Snell, Ensign, USNRF
O. S. Mock, Gunner, USNRF
F. H. Ogle, Pharmacist (T), USN
G. R. Heissel, Act. Pay Clerk (T), USN
J. W. Luce, Act. Pay Clerk (T), USN
H. G. S. Wallace, Commander, USN
J. D. Hashogan, Lieut. Comdr., USNRF
J. L. Begelon, Lieutenant (jg), USNRF
J. L. Nowell, Ensign (T), USN
E. H. Williams, Ensign, USNRF
H. T. Johnson, Asst. Surgeon, USNRF
V. B. Gilman, Asst. Paymaster, USNRF
W. L. Rooney, Asst. Paymaster, USNRF
C. M. Anderson, Lieutenant, USNRF
H. J. Thompson, Lieutenant, USNRF
B. L. Barofsky, Ensign (T), USN
E. D. M. Payne, Ensign (T), USN
T. C. Batdorf, Ensign, USNRF
J. J. Fitzgerald, Ensign, USNRF
G. C. Forrester, Ensign, USNRF
D. A. Freeman, Ensign, USNRF
W. F. Guy, Ensign, USNRF
N. C. Lee, Ensign, USNRF
S. T. Lemson, Ensign, USNRF
P. D. Reynolds, Ensign, USNRF
I. G. Seey, Ensign, USNRF
H. C. F. Wissman, Ensign, USNRF
F. W. Wolf, Ensign, USNRF
J. D. Blackwood, Lieut. (MC), USNRF
L. M. Smith, Lieut. O'g) (MC), USNRF
H. J. Lehman, Lieut, (jg) (MC), USN
F. J. Sullivan, Ensign (PC), USNRF
J. W. Rabbit, Gunner (T), USN
W. G. Bisel, Pharmacist (T),
USN J. G. Sabe, Machinist, USNRF

U. S. S. Agamemnon

(Formerly KAISER WILHELM II) (Transport)

C. B. Morgan, Captain, USN
Wallace Bertholf, Lieut. Comdr., USN

B. R. Ware, Lieutenant, USN
M. Collins, Lieutenant, USN
J. A. Bumette, Lieutenant, USNRF
A. G. Velton, Lieutenant, USNRF
E. Denzler, Lieutenant, (jg) USNRF
C. J. Bell, Lieutenant (jg). USNRF
M. Dumars, Lieutenant (jg), USNRF
W. T. Crowlev, Lieutenant (jg), USNRF
J. J. Parker, Lieutenant (jg), USNRF
W. E. O'Connell, Ensign, USN
J. H. Lawson, Ensign, USN
A. H. Dodge, Surgeon, USN
W. G. Neill, Paymaster, USN
J. N. B. Hill, Asst. Paymaster, USNRF
C. H. Fogg, Gunner, USN
H. P. K. Lyons, Carpenter, USN
G. F. Veth, Machinist, USN
P. R. Abrams, Machinist, USN
W. R. Joiner, Pharmacist, USN
L. H. French, Pharmacist, USN
Frank Maytham, Lieutenant, NNV
J. E. Powell Lieutenant, USNRF
Moses Dumara, Lieutenant (jg), USNRF
C. H. Hermance, Lieut, (jg), USNRF
P. R. Abrams, Ensign, USN
J. J. Enders, Ensign, USN
R. A. Scott, Ensign, USN
W. H. O'Donoghue, Ensign, USNRF
E. P. Nevin, Ensign, USNRF
G. F. Meares, Ensign, USNRF
C. W. Colonna, Asst. Surgeon, USN
R. Heym, Asst. Siu-geon, NNV
J. P. Helman, Asst. Paymaster, USNRF
W. E. Morton, Pay Clerk, USN
S. E. Haddon, Pay Clerk, USN
H. L. Leeb, Ensign, USNRF
E. O'Brien, Ensign, USNRF
L. Lockwood, Asst. Paymaster, USN
K. J. Chvens, Asst. Paymaster, USN
M. L. Pittman, Lieutenant, USNRF
C. G. Muller, Lieutenant, USNRF
E. R. Olmstead, Ensign, USNRF
A. D. Delmer, Ensign, USNRF
S. R. Mackie, Ensign, USNRF
C. K. Etter, Ensign, USNRF
R. D. Team, Asst. Surgeon, USN
E. C. Melton, Asst. Surgeon, USNRF
L. F. Snvder, Dental Surgeon, USN
H. Shortall, Ensign, USNRF
F. C. Sammons, Ensign, USNRF
J. J. Starrow, Ensign, USNRF
F. W. Clements, Ensign, USNRF

O. R. Flagg, Ensign, USNRF
J. A. Biello, P. A. Surgeon, USN
W. E. Meadows, Chaplain, USN

Relief Officers in Order of Reporting on Board

David F. Sellers, Captain, USN
O. E. Grimm, Lieutenant, USN
T. S. Maple, Ensign, USNRF
J. E. Coane, Ensign, USNRF
E. F. Jardine, Ensign, USNRF
H. R. Hobson, Ensign, USNRF
H. R. Baker, Ensign, USNRF
A. E. Conner, Ensign, USNRF
J. E. Doward, Ensign, USNRF
C. H. Whitney. Ensign, USNRF
J. W. Fitzpatrick, Ensign, USNRF
B. T. Campbell, Ensign, USNRF
G. G. Kluber, Ensign, USNRF
A. W. McGinnis, Ensign, USNRF
C. J. Buck, Asst. Paymaster, USNRF
B. H. Micou, Asst. Paymaster, USNRF
J. E. Topliffe, Asst. Paymaster, USNRF
H. J. Mcgin, Pharmacist, USNRF
H. E. Meyers, Boatswain, USNRF
E. F. Ilardine, Lieutenant (jg), USNRF
A. J. Hcnriqucs, Lieut. USNRF
J. J. Orr, Ensign, USNRF
B. L. Lavender, Ensign, USNRF
W. F. Cleveland, Ensign, USNRF
M. S. P. Williams, Ensign, USNRF
Bruce R. Ware, Lt. Comdr. USN
L. Blanchard, Lieutenant, USNRF
W. M.Fleishman, Lieutenant, USNRF
C. K. Blackburn, Lieutenant, USN
F. A. Green, Ensign, USNRF
J. N. Whipple Ensign, USNRF
D. B. Fulton, Ensign, USNRF
B. D. Conant, Ensign, USNRF
A. M. Billings, Ensign, USNRF
J. E. Murphy, Ensign, USNRF
W. F. De Sliding, Ensign, USNRF
F. C. Kukurk, Ensign, USNRF
J. P. Downing, Ensign, USNRF
W. B. Dowie, Ensign, USNRF
J. V. Klemann, Captain, USN
R. H. Blake, Ensign, USNRF
W. E. Batty. Ensign, USNRF
L. Beekman, Ensign, USNRF
L. B. Beatty, Ensign, USNRF
A. B. Bennett, Ensign, USNRF
J. W. Beatty, Ensign, USNRF

T. J. Costello, Ensign (T), USN
C. E. Barnes, Ensign, USNRF

U. S. S. Albany
(Cruiser)

J. J. Raby, Commander, USN
C. S. McWhorter, Lieutenant, USN
I. W. Bobbins, P. A. Surgeon, USN
J. E. Brenner, Lieutenant (jg), USN
C. W. LeRoy, Pay Clerk, USN
C. F. Osborn, Ensign, USN
R, W. Wuest, Lieutenant, USN
R. J. Miller, Lieutenant, USN
M. Hodson, Lieutenant, USN
T. R. Bunting, Gunner (R), USNRF
R. V. Miller, Lieutenant (jg), NNV
A. R. Schofield, Asst. Paymaster, USN
J. R. Sullivan, Ensign, USN
P. R. Taylor, Ensign, USN
T. J. Haffey, Ensign, USN

Relief Officers in Order of Reporting on Board

J. D. Wognum, Pay Clerk (T), USN
W. G. Scott, Carpenter (T), USN
J. Leeming, Ensign, USNRF
A. H. Acorn, Ensign, USNRF
S. S. Yeandle, 2nd Lieutenant, USCG
M. J. Ryan, 2nd Lieutenant, USCG
J. J. Hendren, Asst. Surgeon, USN
V. Peterson, Gunner (O), USNRF
C. H. Crawford, Boatswain, USNRF
F. D. narrower. Ensign (T), USN
W. C. Rickerson, Ensign (T), USN
C. E. Nordhus, Ensign (T), USN
W. W, Funk, Ensign (T) (NE), USN
D. D. Smead, Jr., Ensign (T), USN
C. D. Everingham, Asst. Paymaster, USNRF
T. H. Boyce, Pay Clerk (Act.) USN
W. C. Watts, Commander, USN
M. C. Forster, Ensign (T), USN
G. F. Mentz, Ensign, USN
J. C. Metzel, Ensign, USN
P. Van R. Harris, Lieutenant (jg), USN
W. W. Weld, Lieutenant (jg), USN
C. W. Hanna, Lieutenant (jg), USNRF
J. H. Flagg, Ensign, USNRF
R. W. Collins, Ensign, USNRF
E. B. Cantey, Asst. Paymaster, USN
C. F. Adams, Ensign, USN

R. P. Adair, Ensign, USNRF
John R. Adams, Ensign, USNRF
C. M. Glassmire, Lieut. (MC), USN
H. A. Shepard, Ensign. USNRF
R. T. Weber, Ensign, USNRF
C. L. McCune, Ensign, USN

U. S. S. America
(Transport)

G. C. Day, Captain, USN
F. L. Oliver, Lieut. Comdr., USN
C. M. Peck, Lieut. Comdr., NNV
W. W. Turner, Lieutenant, USN
William Mallett, Lieutenant, USNRF
C. S. Sholes, Lieutenant (jg), USNRF
F. Keene, Lieutenant (jg), USNRF
Lowell Cooper, Ensign, USN
C. A. Lombard, Ensign, USNRF
W. J. Wilkie, Ensign, USNRF
D. McCarthy, Ensign, USNRF
R. W. Ehrhardt, Ensign, NNV
J. D. Manchester, Surgeon, USN
A. H. Deering, Asst. Surgeon, USNRF
G. B. Bloomer, Asst. Paymaster, USNRF
T. D. Loughlan, Asst. Paymaster, USNRF
H. W. Smith, Boatswain, USNRF
C. A. Grove, Boatswain, USNRF
A. Wing, Boatswain, USN
J. J. Madden, Gunner, USN
D. Duffy, Gunner, USN
J. M. McEwen, Gunner, USNRF
L. T. Griffin, Machinist, USNRF
A. H. Hoffman, Machinist, USNRF
M. Bayer, Machinist, USN
T. L. Hannah, Asst. Nav. Constr., USN
W. E. Bassett, Carpenter, USN
C. A. Rowe, Ensign, USN
W. H. McWilliams, Pharmacist, USN
M. E. Huntley, Lieutenant, USNRF
W. L. Ainsworth, Lieutenant, USNRF
A. M. Jones, Asst. Paymaster, USN
R. A. Blair, P. A. Surgeon, NXV
D. G. McRitchie, Paymaster, USN
G. G. Irwin, Asst. Surgeon, USN
J. S. Silvia, Lieutenant (jg), NNV
E. C. O'Shea, Machinist, USNRF
J. M. Lynch, Surgeon, USNRF
R. H. Krepps, Asst. Surgeon, USN
D. Barns, Pay Clerk, USNRF
W. B. Gordon, Lieutenant (jg), USNRF
W, Tillotson, Carpenter, USNRF

C. C. Pendleton, Gunner, USNRF
J. W. McDonald, Lieut, (jg), USNRF
John B. Faison, Ensign, USNRF
Robert A. Farnum, Ensign, USNRF
John F. Killgrew. Ensign, USNRF
Ira H. Meyers, Ensign, USNRF
E. N. Sweitzer, Ensign, USNRF
R. I. Longabaugh, Surgeon, USN
George A. Alden, Asst. Surgeon, USN
W. H. Pilcrantz, Pay Clerk, USNRF
C. H. Price, Lieutenant (jg), USN
J. G. M. Stone, Lieutenant (jg), USN
S. G. Norton, Lieutenant (jg) USN
R. M. Gerth, Ensign, USN
J. McKean, Ensign, USN
A. F. Rodrick, Asst. Surgeon, USNRF
C. V. Van Gassbeck, Asst. Surgeon, USN
L. B. Slickter, Ensign, USNRF
A. J. McDaniel, Pay Clerk, USN
M. M. Leonard, Chaplain, USN
V. B. Allison, Ensign, USNRF
B. T. Barber. Ensign, USNRF
E. D. Baker, Jr., Ensign, USNRF
F. J. Barden, Ensign, USNRF
J. Bartlett, Ensign, USNRF
J. J. Patterson, Ensign, USN
L. C. Parker, Ensign, USN
R. L. Bent, Ensign, USN
A. Anable, Ensign, USNRF
F. H. Bobyshell, Ensign, USN
L. S. Winston, Ensign, USNRF
E. O. Eckdahl, Ensign, USNRF
C. L. Poor, Ensign, USNRF
V. P. Anderson, Ensign, USN
J. J. Fitzgerald, Ensign, USNRF
D. M. Taylor, Ensign, USNRF
G. W. Faber, Ensign, USNRF
A. E. Burroughs, Asst. Paymaster, USNRF

Relief Officers in Order of Reporting on Board

Z. E. Briggs, Captain, USN
F. Spoerr, Lieutenant, USX'RF
R. E. Brush, Lieutenant (jg), USNRF
G. Fried, Lieutenant (jg), USNRF
C. O. Savage, Ensign, USNRF
S. H. Noble, Ensign, USNRF
W. G. Russell, Ensign, USNRF
T. F. Crowther, Ensign, USNRF
E. L. Walter, Lieut, (jg) (MC), USN
J. F. Loba, Lieut, (jg) (PC), USN
W. J. Whelan, Ensign, (PC) USNRF

R. H. Stanly, Pharmacist, USN
K. H. Goss, Pay Clerk, USNRF
G. P. Courtnay, Lieutenant (jg) (MC), USNRF
A. C. Blanding, Lieutenant (jg), USNRF
W. S. Block, Lieutenant (jg), USNRF
W. C. Martin, Lieutenant (jg), USNRF
A. S. Coble, Ensign, USNRF
B. C. Brown, Ensign, USNRF
F. C. Burk, Ensign, USNRF
E. P. Bruch, Ensign, USNRF
J. W. Butterick, Ensign, USNRF
D. T. Smith, Ensign. USNRF
R. Cleeland, Ensign, USNRF
W. W. Hawkes, Lieut. (MC), USNRF
McD. Scott, Lieut. (MC), USNRF
C. C. Pendleton, Gunner, USNRF
Del. M. Young, Gunner, USN
E. Hamilton, Pay Clerk, USNRF

U. S. S. Antigone
(Transport)

J. R. Defrees, Commander, USN
I. C. Bogart, Lieutenant, USN
T. W. Sheridan, Lieutenant, USNRF
H. E. Berg, Lieutenant (jg), USNRF
P. C. Morgan, Lieutenant (jg), USNRF
R. R. Claghorn, Ensign, USN
R. G. Moody, Ensign (T), USN
A. K. Goffe. Ensign (T). USN
A. Eldridge, Ensign (T), USN
F. F. Webster, Ensign (T), USN
A. E. Freed, Ensign (T), USN
F. R. Wilson, Ensign, USNRF
H. C. Curl, Medical Inspector, USN
H. L. Brown, P. A. Surgeon, I'SN
R. W. Clark, P. A. Paymaster, USN
H. F. Baske, Asst. Surgeon, USN
Paul Keller, Asst. Surgeon, USN
A. S. Jewett, Asst. Paymaster, USNRF
D. P.Marting, Asst. Paymaster, USNRF
J. A. Cook, Carpenter, USN
C. A. Werner, Machinist, USNRF
J. M. Thomas, Pay Clerk (T), USN
C, B. Kirkpatrick, Ensign, USNRF
K. C. Mcintosh, P. A. Paymaster, USN
R. C. Adams, Asst. Paymaster, USN
J. F. Halloran, Asst. Paymaster, USN
C. C. Laws, Boatswain, USN
H. E. Humphreys. Pay Clerk (T), USN
R. H. Gibson, Ensign, USNRF
M. Case, Asst. Surgeon, NNV

H. Top, Pharmacist (T), USN
W. J. Bisel, Pharmacist (T), USN
A. C. Dennison, Ensign, USNRF
C. H. Henjes, Ensign, USNRF
F. W. Girdner, Ensign, USNRF
A. Schwartz, Ensign, USNRF
A. R. Gay, Act. Chaplain, USN
H. Brannan, Carpenter (T), USNRF
J. A. Crocker, Lieutenant, USNRF
F. R. Nichols, Lieutenant, USNRF
M. L. Dunn, Ensign (E), USNRF
O. J. Case, Asst. Surgeon, NNV
Daniel Hunt, P. A. Surgeon, USN
Arthur Joachims, Lieutenant, USNRF
O. R. Flagg, Ensign, USNRF
A. J. Chenery, Asst. Surgeon, USNRF
R. Irving, Ensign, USNRF
H. M. Fleetwood, Lieutenant, USNRF
R. D. Beckford, Ensign, USNRF
E. L. Casey, Ensign, USNRF
A. E. Friedman, Ensign, USNRF
J. S. Hanna, Ensign, USNRF
A. C. Haven, Ensign, USNRF
S. W. Higgins, Ensign, USNRF
D. H. Marsh, Ensign, USNRF
P. L. Mather, Ensign, USNRF
S. A. Mead, Ensign, USNRF
1. Schwab, Ensign, USNRF
J. L. Leary, Ensign, USNRF
J. J. Gilham, Midshipman, USN
F. H. Gilmer. Midshipman, USN
T. J. Griffin. Midshipman, USN
W. B. Lower, Asst. Paymaster, USNRF
H. C. Coburn, Lieutenant, USNRF
G. P. Kenney, Lieutenant, USNRF
J. A. Regnier, Dental Surgeon, USNRF
Jeremiah Harris, Asst. Surgeon (T), USN
B. Stuart, Ensign, USNRF
P. S. DeGrouchy, Ensign, USNRF
P. Booth, Ensign, USNRF
T. R. Jones, Ensign, USNRF

Relief Officers in Order of Reporting on Board

Geo. M. Baum, Commander, USN
John Davis, Lieutenant (jg), USNRF
A. H. Dearing Lieutenant (MC) USN
Edward W. Neville Ensign (T), USN
C. L. Hoffman, Ensign (E). USNRF
H. W. Osterhaus, Captain, USN
J. P. Bretherton, Ensign, USNR
J. L. Burt, Ensign, USNRF

R. H. Bowers, Ensign, USNRF
D. J. Brightman, Ensign, USNRF
J. S. Blumenthal, Ensign, USNRF
C. J. Glover, Ensign (PC), USNRF
D. J. Brawner, Gunner (T), USN
Dave Shoemaker, Boatswain (T), USN

U. S. S. Calamares

(Transport)

C. L. Arnold, Commander, USN
E. A. Lichtenstein, Lieut. Comdr., USN
R. S. Parr, Lieutenant, USN
C. L. Jacobsen, Lieutenant, USN
C. F. Hyrne, Lieutenant, USN
C. L. Baker. Lieutenant, USNRF
A. M. Austin, Lieutenant (jg), USNRF
R. W. Dearborn, Lieut, (jg), USNRF
E. Rang, Lieutenant (jg), USNRF
W. Harrington, Lieut, (jg), USNRF
T. Mulholland, Lieut, (jg), USNRF
C. D. Draper, Lieutenant (jg), USNRF
J.W. S. Smith, Lieutenant (jg), USNRF
P. S. Hirst, Ensign, USNRF
A. E. Nelson, Ensign, USNRF
G. O. Gustafson, Ensign, USNRF
P. Rix, Ensign, USNRF
W. A. Hawke, P. A. Surgeon, USNRF
L. G. Jordan, Asst. Surgeon, USN
H. L. Howell Asst. Surgeon, USNRF
C. G. Warfield, Asst. Paymaster, USN
A. Eldred, Asst. Paymaster, USNRF
A. E. Chase, Asst. Paymaster, USNRF
M. J. Hannafein, Boatswain, USN
V. Peterson, Gunner, USN
R. S. Lunney, Gunner, USN
V. H. Richards, Machinist, USN
W. E. Acton, Jr., Carpenter, USN
N. W. Parks, Pharmacist, USN
C. A. Cameron, Pav Clerk, USN
F. W. S. Dean, Medical Inspector, USN
T. A. Clark, Carpenter, USN
S. Cochran, Lieut. Commander, USN
J. W. Baldwin, Lieutenant, USNRF
E. V. Brewer, Ensign, USNRF
R. R. Beatty, Ensign, USNRF
P. J. Pond, Ensign, USNRF
R. M. Bourne, Ensign, USNRF
S. F. Boyd, Ensign, USNRF
L. H. Ackerman, Lieutenant, USNRF
J. A. Regnier, Dental Surgeon, USNRF
M. J. Hannafin, Boatswain (T), USN

C. H. Christainsen, Ensign, USNRF
E. S. Templeton, Ensign, USNRF
R. M. White, Ensign, USNRF
H. K. Wilson, Ensign, USNRF
A. F. Anglemyer, Ensign, USNRF
L. H. Bodman, Ensign, USNRF
W. B. Boise, Ensign, USNRF
L H. Jones, Ensign, USNRF
Thomas Blau, Lieut. Comdr., USNRF
H. T. Mitchell, Lieutenant(jg), USNRF
E. G. Bachman, Lieut, (jg), USNRF
J. J. Cooney, Ensign, USN
A. P. Croucher, Ensign, USN
A. G. Crafts, Ensign, USN
N. R. Copeland, Ensign, USN
C. T. Dimmitt, Ensign, USN
T. M. Duff, Ensign, USN
H. C. Keil, Machinist, USNRF
C. E. Tupper, Machinist, USNRF
J. A. Kelley, Machinist, USNRF
C. C. Ammeron, Lieut. (MC), USNRF
E. W. Fenton, Lieut. (PC) (jg) USNRF
D. M. Jones, Gunner (E), USNRF
C. H. Hebble, Machinist, USN
C. H. Hammond, Machinist, USNRF
M. L. Fortier, Machinist, USNRF

U. S. S. Charleston

(Cruiser)
Flagship of Rear-Admiral H. P. Jones, USN

E. H. Campbell, Commander, USN
Robert Henderson, Lieut. Comdr., USN
C. W. Crosse, Lieutenant, USN
S. F. Heim, Lieutenant, USN
K. H. Donavin, Lieutenant, USN
G. E. Brandt, Lieutenant, USN
T. S. King, Lieutenant (jg), USN
H. S. Keep, Lieutenant (jg), USN
H. W. Hoyt, Lieutenant (jg), USN
G. F. Neiley, Lieutenant (jg), USN
J. H. Smith, Lieutenant, USN
W. D. Bungert, Ensign, USN
H. L. Vickery, Ensign, USN
J. Irwin, Paymaster, USN
J. MacIntyre, Ensign, USN
J. P. Richter, Ensign, USN
F. J. "Wilson, Asst. Nav. Constr., USN
D. C. Beach, Chief Machinist, USN

W. A. Martin, Boatswain, USN
G. Kleinsmith, Gunner, USN
H. H. Bloxham, Pay Clerk, USN
W. A. Brams, Asst. Surgeon, USN
A. E. Stover, Lieutenant (jg), NNV
W. A. Taylor, Lieutenant (jg), NNV
C. C. Jackson, Lieutenant (jg), NNV
S. C. Williams, Lieutenant (jg), NNV
W. S. Forrest, Lieutenant O'g). NNV
G. L. Gens, Ensign, NNV
E. R. Bussler, Ensign, USNRF
R. E. Allen, Ensign, USNRF
H. B. Converse, Ensign, USNRF
D. L. Noves, Ensign, USNRF
R. W. Dolton, Ensign, USNRF
G. L. Heyer, Ensign, USNRF
G. Nolan, Ensign (NM), USNRF
J. B. Sasse, Machinist, USN
M. J. Stubbs, Asst. Paymaster, USN
A. B. Clark, Asst. Paymaster, USN
C. F. Manley, Carpenter, USN
Louis Uttendorfcr, Boatswain, USN
F. W. Nehls, Boatswain, USNRF
O. A. Ilelraerichs, Boatswain, USNRF
O. Masscn, Boatswain, USNRF

F. T. Evans. Lieut. Commander, USN
E. M. Mullen, Asst. Surgeon, USN
M. L. Carr, Asst. Surgeon, USN
W. F. Olson, Ensign, USNRF
L. E. Burwell, Ensign, USNRF
M. Erickson, Carpenter, USN
J. E. Ohlson, Carpenter, USN
J. W. Lucas, Gunner (R), USNRF
M. E. Pope, Machinist, USN
E. M. Thompson, Machinist, USN
O. Clark. Lieutenant (jg), USN
J. A. Sternberg, Lieutenant (jg), USN
E. P. Sauer, Lieutenant (jg), USN
C. H. White, Ensign, USNRF
P. W. Richard, Ensign, USNRF
G. T. Ellis, Ensign, USNRF
M. F. Werner, Gunner (T), USN
K. Farnum, Act. Pay Clerk, USN
W. J. Rodgers, Dental Surgeon, USNRF
E. C. Long, Captain, USMC
W. M. Marshall, Captain, USMC
J. F. Hines, Captain, USN
W. H. Shea, 1st Lieutenant, USCG
S. P. Sears, Ensign, USNRF

E. L. McSheehy, Lieut. Comdr., USN
J. W. Chapman. Ensign (T), USN
R. H. Finlay. Ensign (T), USN
B. A. Grimball, Ensign (T), USN
W. P. McCoy, Ensign (T), USN
W. M. Akin, Ensign (T), USN
D. H. McCoy. Ensign USNRF
R. A. McCloud, Ensign, USNRF
E. R. Maillette, Act. Pay Clerk, USN
W. I. Greth, 2nd Lieutenant, USMC
L. A. Odlin, P. A. Paymaster, USN
C. G. Holland, P. A. Paymaster, USN
R. E. Kline, Pay Clerk, USNRF
E. A. A. Gendreau, Asst. Surgeon, USN
F. J. Hurney, Act. Chaplain, USN
E. Danielson, Ensign (T), USN
T. H. Ross, Ensign, USNRF
C. C. Jackson, Lieutenant (jg). USNRF
R. P. Bell, Asst. Surgeon, USNRF
C. P. Williams, Ensign, USNRF
W. L. Littlefield, Captain, USN
E. D. Washburn, Commander, USN
F. C. Allen, 1st Lieutenant. USCG
R. V. Ahlstrom, Ensign (T), USN
C. S. Allen, Ensign (T), USN
A. D. Alexander, Ensign (T), USN
E. M. Alexander, Ensign (T), USN
W. Atherton, Ensign (T). USN
W. E. Andrews, Ensign (T), USN
E. W. Bacon, Ensign (T), USN
M. T. Clement, Lieutenant (MC), USN

U. S. S. Chattanooga
(Cruiser)

Arthur MacArthur, Commander. USN
Ellis Lando, Lieutenant. USN
Alexander Macomb, Lieut, (jg), USN
A. L. Morgan, Lieutenant (jg), USN
Edward Breed, Ensign. USN
E. G. Herzinger, Ensign. USN

Relief Officers in Order of Reporting on Board

C. L. Hansen. Ensign. USN
Washington Bogardus, Ensign, NNV
R. B. McEwan. Ensign, NNV
Morrison B. Orr, Ensign, NNV
William T. Brown, Ensign, NNV
W. K. Blair, Ensign, NNV
Parker C. Hatch, Ensign, NNV
W. H. French, Ensign, USNRF

Harold F. Fultz, Ensign, USNRF
Sydney W. Ford, Ensign. USNRF
R. T. Mahon, Asst. Paymaster. USN
Morris A. Jacobs, Lieut, (jg), USNRF
Fred. A. Zscheuschler. 2nd Lieut., USCG
Earl F. Chandler, Ensign, USNRF
Eugene R. Black, Ensign, USNRF
P. Seaman Bleecker. Ensign, USNRF
Kenneth L. Coontz, Ensign, USN
Ernest H. Barber, P. A. Paymaster, USN
Roland R. Gasser, Asst. Surgeon, USN
Roy E. Smith, Pay Clerk, USN
E. D. Wallridge. Ensign, USN
Theo. F. C. Walker, Ensign, USN
Kenneth H. Stetson. Ensign (T), USN
Ralph A. Ofstie, Midshipman, USN
Charles A. Nicholson, Ensign, USN
George N. Herring, Asst. Surgeon. USN
Harry K. Cage, Commander, USN
Chester K. Harrison, Ensign, USNRF
H. V. Adams, Ensign. USNRF
Louis Apfelbaum, Ensign. USNRF
H. Carson. Ensign, USNRF
M. B. Savage, Ensign, USNRF
Henry K. Barwick, Ensign (T). USN
Roland S. Bailey, Ensign (T). USN
William Bailey, Ensign (T), USN
Charles C. Wolcott, Lieut. (MC). USN
Charles B. Forrest, Pay Clerk, USN

U. S. S. Chicago
(Cruiser)

H. G. Sparrow, Captain, USN
R. W. Kessler, Commander, USN
L. W. Hesselman, Lieut. Comdr., USN
W. G. Hodgson, Lieut. Comdr., USN
H, E. WTiite, Lieutenant, USN
G. R. Veed, Lieutenant, USN
W. R. Cole, Lieutenant (jg), USNRF
H. G. Simonds, Ensign, USNRF
E. C. Welch, Ensign, USNRF
J. M. Quinlan, Lieutenant (jg), USN
R. S. Babcock, Ensign, USN
G. M. Kennedy, Lieutenant (MC), USN
B. F. McDonald, Lieut. (MC), USN
H. B. Teegarden, Lieut. (PC), USN
E. J. Byrnes, Machinist, USN
R. Odening, Machinist, USN
J. V. Thomas, Machinist, USNRF
W. J. Schmidt, Carpenter, USN

O. F. Bvrd, Act. Pay Clerk, USN
P. J. Fleming, Act. Pay Clerk. USN
J. E. Wood, Act. Pay Clerk, USN

U. S. S. Cleveland
(Cruiser)

John F. Hines, Commander, USN
Carl C. Krakow, Lieutenant, USN
William E. Baughman, Lieut, (jg), USN
Franz B. Melendy, Lieut, (jg), USN
William F. Roehl, Lieut, (jg), USN
William D.Sullivan, Ensign, USN
John E. Reinburg, Jr., Ensign, USN
William W. Schott, Ensign, USN
Thomas L. Sprague, Ensign, USN
Earl E. Stone, Ensign, USN
Donald A. Green, Ensign, USN
Grover C. Wilson, Asst. Surgeon, USN
Arthur H. Mayo, P. A. Paymaster, USN
Wilson S. Hullfish, Asst. Paymaster, USN

Relief Officers in Order of Reporting on Board

Ray E. Ames, Pay Clerk, USN
K. M. Bennett, Commander, USN
J. S. Brayton, Ensign, USNRF
J. O. Burgwin, Ensign, USNRF
H. Butler. Ensign, USNRF
C. Chapman, Ensign, USNRF
W. B. Hanley, Act. Pay Clerk, USN
P. M. Lund, Act. Gunner, USNRF
C. R. Miller, Commander, USN
W. J. McDonald, Ensign, NNV
G. W. McKean, 3rd Lieutenant, USCG
C. W. Taylor, Gunner (E), USNRF
A. O. Mang, Machinist, USN
Douglas S. Moore, Ensign (T), USN
Kenneth C. Root, Asst. Paymaster, USN
W. T. Smart, Boatswain, USN
Thomas W. Allen, Ensign (T), USN
A. O. Gies, Ensign (T), USN
H. R. Homer, Ensign (T), USN
R. B. Schaal, Ensign (T), USN
J. S. Albany, Ensign (T), USN
S. Fried, Ensign (T), USN
S. E. Howes, Ensign (T), USN
A. Grove, Boatswain (T), USN
H. M. Martin, Ensign, USN
S. G. Lamb, Ensign, USN
W. J. Corcoran, Asst. Surgeon, USN
J. J. Hyland, Captain, USN

H. L. Carter, Ensign, USNRF
W. J. Stultz, Ensign, USNRF
O. W. Blackett, Ensign (T), USN
G. R. Bedenkopp, Ensign (T). USN
F. H. Baxter, Ensign, USNRF
T. S. Woods, Ensign, USNRF
R. S. Lawson, Ensign, USNRF
L. O. Crocker, Ensign, USNRF
L. V. Lizars, Ensign, USNRF
M. S. Bender, Lieutenant (MC), USN

U. S. S. Columbia
(Cruiser)

F. B. Upham, Captain, USN
C. C. Moses, Lieut. Commander, USN
Weyman P. Beehler, Lieutenant, USN
William H. Porter, Lieutenant, USN
E. R. McClung, Lieutenant (js). USN
C. W. McNair, Lieutenant (jg), USN
L. W. Clarke, Lieutenant (jg) USN
J. D. Ross, Lieutenant Q'g). NNV
J. W. Fowler, Ensign, USN
O. F,. Grimm, Ensign, USN
L. L. Habrylewicz, Ensign, USN
C. W. Styer, Ensign, USN
H. W. Brown, Ensign, NNV
H. S. Alden, Ensign, NNV
A. R. Gilman, Ensign, USNRF
J. P. Cozzens, Ensign, USNRF
S. J. Meeker, Ensign, USNRF
W. S. Forsyth, Ensign, USNRF
A. S. Neilson, Ensign, USNRF
H. K. McHarg, Ensign, USNRF
O. J. Mink, Surgeon, USN
J. H. Chambers, Asst. Surgeon, USN
H, B. Ransdell, P. A. Paymaster, USN
W. W. Elder, Act. Chaplain, USN
F. E. Chester, Ensign (T). USN
G. Crofton, Ensign (T), USN
Niels Drustrup, Gunner, USN
M. Dickinson, Gunner, USN
P. L. Elkins, Machinist, USN
R. G. McClure, Machinist, USN
A. Tucker, Asst. Constr., USN
J. G. Stanton, Pay Clerk, USN

Relief Officers in Order of Reporting on Board

Harmon, Ensign, USNRF
G. G. Jones, Ensign, USNRF
M. L. Royar, Asst. Paymaster, USN

F. Ellison, Boatswain (T), USN
H. E. Myers, Boatswain (T), USN
W. A. Doty, Machinist (T), USN
F. L. Austin, 1st Lieutenant, USCG
R. H. Jones, Lieutenant (jg), USN
M. Comstock, Lieutenant (jg), USN
H. Evans, Ensign (T), USN
T. R. Jones, Ensign, USNRF
C. K. Wallace, Ensign (T), USN
B. F. Schauffler, Ensign (T), USN
T. T. Hassell, Ensign (T), USN
C. J. Koehler, Ensign (T), USN
E. Denton, Ensign (T), USN
W. E. Skinner, Ensign, USNRF
H. S. Dewey, Ensign, USNRF
H. A. Mitchell, Ensign, USNRF
C. H. Grove, Lieutenant (jg), USNRF
A. B. McKeel. Chief Carpenter, USN
M. C. McCray, Pay Clerk, USN
W. S. Rockwell, Pay Clerk, USN
G. Poggi, Pay Clerk, USNRF
W. T. Crone, Ensign (T), USN
B. Richcreek, Boatswain (T), USN
O. B. Bennet. Pay Clerk, USN
R. L. Hicks, Ensign, USN
J. F. Halloway, Ensign, USN
R. C. Graves, Ensign, USNRF
T. R. Hicks, Ensign, USNRF
J. McK. Spears, Ensign, USNRF
C. B. Sherman, Ensign, USNRF
E. I. Taylor, Ensign, USNRF
C. W. O. Goodwin, Ensign, USNRF
John Gallagher, Lieut, (jg) (TM), USN
Earl C. Carr, Asst. Surgeon, USN
H. L. Brinser, Captain, USN
J. J. London, Commander, USN
W. C. Bowne, Ensign, USNRF
C. R. Brick, Ensign, USNRF
B. S. Blanchard, Ensign, USNRF
K. A. Burger, Ensign, USNRF
H. H. Brakeley, Ensign, USNRF
C. B. Sheridan, Ensign (T),
USN E. I. Taylor, Ensign, USNRF
C. W. O. Goodwin, Ensign, USNRF
S. S. Posnaugh, Boatswain (T),USN
J. S. Van Winkel, Lieutenant (jg) (MC), USN
K. A. Burgess, Ensign, USNRF
J. T. Wrightson, Ensign, USNRF
Jos. Boudette, Ensign, USNRF

U. S. S. Covington
(Transport)

R. D. Hasbrouck, Captain, USN
C. L. Arnold, Lieut. Commander, USN
C. S. Gillette, Lieutenant, USN
W. J. Hine, Paymaster, USN
P. H. Stibbens, P. A. Surgeon, USN
H. J. White, Ensign, USN
F. Rasmussen, Chief Bosn., USN
R. M. Huggard, Chief Machinist, USN
C. V. Kane, Gunner, USN
R. F. MacDonald, Gunner, USN
R. P. Roberson, Carpenter, USN
H. C. Roe, Pharmacist, USN
G. Wilshire, Ensign, NNV
G. T. January, Lieutenant, USNRF
J. O. Porter, Lieutenant, USNRF
R. M. Packer, Lieutenant (jg), USNRF
C. B. Pengar, Lieutenant (jg), USNRF
N. M. Goodwin, Lieut, (jg), USNRF
B. C. Edwards, Lieutenant (jg), USNRF
A. S. Whitehead, Lieut, (jg), USNRF
J. A. Johnson, Lieutenant (jg), USNRF
A. W. Gould, Lieutenant (jg), USNRF
E. C. Steinhart, Ensign, USNRF
R. S. Smead, Ensign, USNRF
D. E. Mason, Ensign. USNRF
W. E. Caddigan, Ensign, USNRF
A. M. Austin, Ensign, USNRF
W. H. Gregg, Ensign, USNRF
R. J. Routledge. Ensign, USNRF
N. C. Rubinsky, Asst. Surgeon, USNRF
W. J. Pennell, Asst. Surgeon. USNRF
J. S. Hill, Lieutenant (jg), USN
A. C. Pettingill, Asst. Paymaster, USNRF
Frederick Kidston, Boatswain (T), USN
S. S. Armstrong, Asst. Paymaster, USN
R. L. Steketee, Asst. Paymaster, USN
J. McDonald, Lieutenant (jg), USNRF
W. B. King, Ensign, USNRF
C. Foose, Ensign, USNRF
M. F. Gate, Ensign, USNRF
P. G. Bertelson, Ensign, USNRF
G. C. Fowler, Dental Surgeon, USN
P. L. Mitchell, Chaplain (A), USN

Relief Officers in Order of Reporting on Board

R. H. Sloan, Lieutenant, USNRF
J. N. Rasmussen, Lieut, (jg), USNRF

T. H. Parsons, Ensign, USNRF
M. H. Sicard, Asst. Surgeon, USNRF
Guy Whitlock, Commander, USN
A. Ohmer, Carpenter, USN
W. M. Fleischman, Lieutenant, USNRF
R. Williams, Ensign, USNRF
E. P. Story, Ensign, USNRF
O. G. Dale, Ensign, USNRF
L. G. Beattv, Ensign, USNRF
H. K. Nickell, Ensign, USNRF
L. A. Lehmair, Ensign, USNRF

U. S. S. De Kalb

(Cruiser)

W. R. Gherardi, Commander, USN
A. S. Wadsworth, Lieutenant, USN
B. A. Strait, Lieutenant, USN
A. R. Simpson, Lieutenant (jg), USN
T. Bruggey, Lieutenant, USN
T. W. Rudderow, Lieutenant, NNV
F. J. Patton, Surgeon, NNV
I. D. Coyle, P. A. Paymaster, USN
J. B. Spencer, Asst. Surgeon, USNRF
R. B. Marble, Jr., Lieut, (jg), USNRF
L. B. Bernheim, Lieut, (jg), USNRF
M. B. Sullivan, Lieut. Gg). USNRF
W. L. Ainsworth, Lieut, (jg), USN
H. B. Howell, Lieut, (jg), USNRF
J. S. Roberts, Ensign, USN
J. S. Burford, Ensign, USNRF
E. N. Fisher, Gunner, USN
C. A. Dannenmann, Boatswain, USN
F. Johnson, Boatswain, USNRF
T. W. Jenkins, Machinist, USNRF
O. D. Parker, Machinist, USN
W. Neidert, Carpenter, USN
E. M. Cronin, Pay Clerk, USN
L. F. Randall, Pay Clerk, USN
M. D. Yokes, Asst. Surgeon, USNRF
H. J. Gosselin, Pay Clerk, USN
J. A. Rittmayer. Pay Clerk, USN
J. A. Alger, Lieutenant, USCG
W. S. Condict, Asst. Surgeon, USNRF
W. L. Kimball, Ensign (T), USN
W. E. Chase, Jr., Ensign (T). USN
E. M. McIlvain, Ensign (T), USN
H. J. Benzing, Lieutenant (jg), USNRF
R. B. Taylor, Ensign, USNRF
H. Casselberry, Ensign, USNRF
V. A. La Barge, Ensign, USNRF
R. A. Jenkins, Ensign, USNRF

M. C. Bird. Ensign, USNRF
J. F. Giiffin, Ensign, USNRF
J. A. Judy, Asst. Surgeon, USNRF

Relief Officers in Order of Reporting on Board

Wm. Baggaley, Lieut. Comdr., USN
T. Bretherton, Asst. Paymaster, USNRF
J. Bona, Gunner (T), USN
L. M. Burt, Gunner (R), USNRF
H. Liebweg, Gunner (T), USN
J. B. Carrol, Bosn., USN
R. IL Nexsen Lieutenant, USNRF
A. P. Spencer, Ensign (T), USN
F. Wurster, Ensign, USNRF
C. D. Wardner, Ensign, USNRF
J. M. Bright, Ensign, USNRF
H. A. Topp. Ensign, USNRF
E. L. Casey, Ensign, USNRF
J, O, Huse, Ensign, USNRF
F. H. Nelson, Ensign, USNRF
J. A. Omer, P. A. Surgeon, USN
W. R. Morton, Asst. Paymaster (T), USN M.
M. P. Kane, Dental Surgeon, USNRF
L. M. Overstreet, Captain, USN
L. L. Walker, Ensign, USNRF
J. M. Dennis, Ensign, USNRF
E. C. Dale, Ensign, USNRF
W. K. Haythorne, Lieutenant, USNRF
F. P. Jones, Dental Surgeon, USNRF
R. B. Rogers, Lieut, (jg), USN (MC)
E. H. Dodd, Captain, USN
E. T. Cook, Ensign, USNRF
C. R. Bradley. Ensign, USNRF
C. R. Randall, Ensign, USNRF
R. J. H. Powell. Ensign, USNRF
W. L. Jones, Ensign, USNRF
N. H. Chase. Ensign, USNRF
L. A. Hill, Ensign. USNRF
L. S. Tailer, Ensign, USNRF
F. Hopkins, Ensign, USNRF
G. I. Murry, Ensign, USNRF

U. S. S. Denver

(Cruiser)

Amon Bronson, Jr., Commander, USN
Newton H. White, Jr., Lieutenant, USN
Robert H. Skelton, Lieut, (jg), USN
William H. O'Brien, Jr., Lieutenant (jg), USN
Frank S. Bloomfield, Lieut, (jg), NNV

John F. Bates, Ensign, USN
Martin B. Stonestreet, Ensign, USN
Leon F. Brown, Ensign, USN
Seldon B. Kennedy, 1st Lieut., USMC
Edward C. Little, P. A. Paymaster, USNRF
Claude W. Carr. P. A. Surgeon, USN
George B. Tyler, Asst. Surgeon, MRC
John W. Rowe, Midshipman, USN
Hayden H. Smith, Midshipman, USN
Ford L. Wilkinson, Jr., Midshipman, USN
Arthur G. King, Pay Clerk, USN

Relief Officers in Order of Reporting on Board

R. F. Herrick, Jr., Ensign, USNRF
J. Hemphill, Ensign, USNRF
A. G. Gennert, Ensign, USNRF
Vi. W. Grace, Ensign, USNRF
Arthur L. Kams, Boatswain, USN
Charles Horsted, Electr. Gunner, USN
George W. Williams, Machinist, USN
Jesse Robertson, Pay Clerk USN
J. G. Payne, Lieutenant (jg), NNV
D. P. Marvin, 2nd Lieutenant, USCG
Edmund Ocumpaugh III, Ensign, USN
Leicester S. Pettit, Ensign, USN
Bruce D. Bromley, Ensign, USNRF
Thomas C. Nicholls, Jr., Ensign, USN
Romaine Hathaway, Pay Clerk, USN
Edward B. Fenner, Commander, USN
J. D. Lowry, Jr., Ensign, USN
Paul F. Lee, Ensign, USN
Mulford M. Stewart, Gunner, USN
Layton B. Carpenter, Ensign, USNRF
Edward M. Chase, Ensign, USNRF
David Crow, Ensign, USNRF
John F. MacDonald, Ensign, USNRF
Nobart W. Thompson, Jr., Ensign, USNRF Samuel M. Beath, Ensign, USNRF
Delbert M. Small, Boatswain, USNRF
George E. Davis, Carpenter, USNRF
John J. Bradley, Ensign, USNRF
William W. Deal, Ensign, USNRF
O. R. Doerr, Ensign, USNRF
T. H. Cherry, Lieut, (jg) (MC), USNRF
James D. Moore, Lieut. Comdr., USN
Martin B. Stonestreet, Lieut., USN

U. S. S. Des Moines
(Cruiser)

J. R. Y. Blakeley, Commander, USN

E. A. Wolleson, Lieutenant (jg), USN
C. A. Bailey, Lieutenant (jg), USN
H. B. Grow, Lieutenant (jg), USN
W. D. Snyder, Ensign, USN
F. D. Wagner, Ensign, USN
O. B. Duncan, Ensign, NNV
R. T. Gallenmore, Ensign, USN
A. F. France, Ensign. USN
E. J. Kidder, Ensign, USN
C. C. Copp, P. A. Paymaster, USN
W. A. Stoops, Asst. Surgeon, USN
E. H. Gale, Pay Clerk, USN

Relief Officers in Order of Reporting on Board

D. W. Wurtsbaugh, Commander, USN
G. T. Jarvis, 2nd, Ensign, USNRF
R. D. Keyes, Ensign, USNRF
P. C. Kauffmann, Ensign, USNRF
J. T. Scully, Ensign, USNRF
H. D. Stillman, Asst. Paymaster, USN
W. T. McMahon, Electr. Gunner, USN
R. F. Streitz, Machinist, USN
W. L. Calloway, Pay Clerk, USN
J. Trebes, Jr., 3rd Lieutenant, USCG
G. W. Haynes, Ensign (T) (NE). USN
P. J. Drake, Ensign, USN
E. H. Ellison, Ensign, USN
J. B. Burnham, Ensign, USN
R. Williams, Ensign, USN
J. P. Gallagher, Pay Clerk, USN
W. O. Skelton, Pay Clerk, USN
L. C. Thyson, Lieutenant (MC), USN
B. Burns. Ensign, USNRF
O. Reran, Ensign, USN
F. P. H. Ackers, Ensign, USNRF
H. K. Adams, Ensign, USNRF
D. M. McGurl, Ensign, USN
W. P. Martin, Ensign, USN
G. D. Howell, Jr., Ensign, USN
A. B. Pedin, Ensign, USNRF
Z. H. Madison, Captain, USN
Louis Polon, Lieutenant (MC), USNRF
H. G. Fuller, Lieutenant, USN
S. J. Burris, Jr., Ensign, USNRF
F. H. Button, Ensign, USNRF
A. C. Dunn, Ensign, USNRF
F. B. Earhart, Ensign, USNRF
W. E. Dunkum, Machinist, USNRF
L. B. Hubbcll, Ensign (T), USN
W. C. Preston, Lieutenant (PC), USN

U. S. S. Finland

(Transport)

W. J. Giles, Commander, USN
A. W. Atkins, Lieut. Commander, USN
J. B. Hill, Lieut. Commander. USNRF
R. F. Skylstead, Lieutenant, USN
John Jenson, Lieutenant, USNRF
C, F. Rogers, Lieutenant, USNRF
A. Jochimsen, Lieutenant, USNRF
G, H. Gaskin, Lieutenant, USNRF
John Muir, Lieutenant(jg) (E), USNRF
W. J. Rague, Lieutenant (jg), USNRF
J. P. Dempsey, Lieutenant (jg), USN
H. O. Carter, Lieut. (jg) (E), USNRF
T. Mathison, Lieutenant (jg), USNRF
Edward Griffin, Ensign, USNRF
J. W. Percy, Ensign, USNRF
C. B. Forbes, Ensign, USNRF
C. R. Streets, Ensign, USNRF
N. W. Nicol, Ensign, USNRF
E. C. Tew, Ensign, USNRF
A. J. M. Grant, Ensign, USNRF
S. Hansen, Ensign, USNRF
G. N. Maynard, Ensign, USNRF
J. M. Burke, Ensign, USNRF
J. L. Shotwell, Ensign (E), USNRF
H. R. Hermeseh, Surgeon, USN
S. C. Strauss, Asst. Surgeon, NNV
J. W. White, Asst. Surgeon, USN
T. S. Wylly, Asst. Paymaster, USN
F. F. Fulton, Asst. Paymaster, USN
Frederick Heim, Boatswain, USN
J. W. Agnew, Gunner (E), USN
W. F. Fitzpatrick, Machinist, USN
L. M. Pierce, Carpenter, USN
E. A. Thiele, Pay Clerk, USN
C. Buskeley, Pay Clerk, USN
A. J. Larson, Pharmacist, USN
J. C. Goff, Machinist, USN

Relief Officers in Order of Reporting on Board

B. R. Lyon, Asst. Surgeon, USN
G. E. Ernst, Ensign, USN
W. F. Morrison, Ensign, USNRF
W. F. Laughlin, Ensign, USNRF
P. O. McDonough, Lieut, (jg), USNRF
M. A. Cox, Lieutenant (jg), USNRF
L. B. O'Shaughnessey, Asst. Paymaster, USNRF
L. B. Cummings, Lieutenant, USNRF

G. S. Hawthorn, Dental Surg., USNRF
R. S. Hotz, Ensign, USNRF
D. E. Craig, Gunner (T), USNRF
P. D. Miller, Ensign, USNRF
C. E. Fortlage, Ensign, USNRF
J. A. Fields, Ensign, USNRF
Harvey F. Hambur, Ensign, USNRF
Geo. W. Stutt, Lieutenant (jg), USNRF
L. C. Krusen, Ensign, USNRF
E. E. Woodland, P. A. Surgeon, USN
Chas. Blount, Ensign. USNRF
Joshua Baker, Ensign, USNRF
W. F. Russel, Lieut, (jg), USNRF
R. B. Coffman, Lieut. Comdr., USN
C. E. Skeen, Lieut, (jg), USNRF
R. I. Dillenback, Lieut, (jg). USNRF
J. C. Fishburn, Ensign (T), USN
C. W. Files, Ensign, USNRF
W. C. Holmes, Ensign (T) (E), USN
R. J. Home, Ensign (T) (E), USN
J. M. Foster, Ensign (D) (T), USN
J. A. Dorgan, Lieutenant (jg), USNRF
J. A. Rasraussen, Lieut, (jg), USNRF
D. E. Walso, Ensign, USNRF
L. S. Ferdon, Ensign, USNRF
E. J. Chapman, Ensign. USNRF

U. S. S. Frederick

(Cruiser)

W. C. Cole, Captain, USN
R. S. Culp, Lieut. Commander, USN
J. J. Manning, Lieut. Commander, USN
J. Wilkes, Lieutenant (jg), USN
I. Parker, Lieutenant (jg), USN
J. A. Scott, Lieutenant (jg), USN
E. M. Hacker, Paymaster, USN
C. B. Munger, Surgeon, USN
A. Rettig, Chief Boatswain, USN
P. J. Gundlach, Ensign (T), USN
R. Semple, Ensign (T), USN
G. C. Smith, Ensign (T), USN
R. R. Clarke, Asst. Nav. Constr. (T), USN
W. D. Dadd, Ensign (T), USN
J. A. Rogers, Ensign (T), USN
J. P. Jackson, Commander, USN
O. P. Oraker, Ensign (T), USN
I. Lehrfeld, Lieutenant (jg), USN
W. S. B. Claude, Lieutenant (jg), USN
A. A. Garcelon, Jr., Lieut. Commander, USN
W. R. Cobb, Lieutenant (jg), NNV
T. C. Pounds, Asst. Surgeon, NNV

S. R. Siebert, Ensign, NNV
H. S. Haynes, Lieutenant (jg), NNV
W. B. Allison, Lieut. Comdr., USN
R. A. Silent, Lieutenant, NNV
W. E. Hubbard, Lieutenant, NNV
A. M. Baldwin, Ensign, NNV
C. B. Tillotson, Ensign, NNV
R. W. Byrns, Act. Pay Clerk, USN
R. R. Thompson, Act. Pay Clerk, USN
M. B. Byington, Ensign, USN
H. K. Leventon, Ensign, USN
J. O. Plonk, Ensign, USN
J. H. Gowan, Radio Gunner (T), USN
O. L. de Vasconcellos (Segundo Teni
ente Brazilian Navy)
M. E. Rothenberg, Radio Gunner (T), USN

Relief Officers in Order of Reporting on Board

W. S. MacKay, Radio Gunner (T), USN
J. Richardson, Machinist (T), USN
G. Glodzei, Ord. Gunner (T). USN
V. E. Stack, Captain, USMC
E. O. Ames, Captain, USMC
E. C. Corwin, Carpenter (T), USN
C. R. Jacobsen, Ensign, NNV
A. F, List, Lieutenant (jg), USNRF
P. G. McKinlay, Lieut, (jg), USNRF
T. J. Parker, Ord. Gunner (T), USN
W. R. La Motte, Lieut, (jg), USNRF
S. S.Gant, Pharmacist (T), USN
E. Richison, Asst. Surgeon, USN
G. S. Townsend, 2nd Lieut., USMC
J. O. Holcomb, Ensign, USNRF
O. McW. Richardson, Ensign, USNRF
G. W. Palmer, Ensign, USNRF
F. L. McNally, Ensign, USNRF
R. T. Jackson, Ensign, USNRF
E. R. Hill, Ensign, USNRF
W. L. Ainsworth, Lieutenant, USN
S. C. Rowan, Commander, USN
J. P. Dalton. Lieutenant, USN
W. F. Roe, Pay Clerk, USNRF
O. L. Brewington, Ensign (T), USN
L. King, Boatswain (T), USN
M, P. Kane, Dental Surgeon, USNRF
C. M. Dixon, Ensign, USNRF
H. T. Greenwood, Ensign, USNRF
H. M. C. Hewson, Ensign, USNRF
T. W. Stein, Ensign, USNRF
E. F. Karges, Ensign, USNRF
H. M. Howett, Ensign, USNRF

J. A. Meeker, Gunner (T), USN
D. S. Robinson, Chaplain, USN
R. S. Jones, Machinist, USN
B. E. Blossei, Gunner (T), USN
H. L. MacBride, Midshipman, USN
H. Marienhoff, Midshipman, USN
W. E. Miller, Midshipman, USN
De L. Mills, Midshipman, USN
C. D. Humphrey, Midshipman, USN
C. L. Hutton, Midshipman, USN
J. V. McElduff, Midshipman, USN
D. A. McMillan, Midshipman, USN
C. B. McVay, 3rd, Midshipman, USN
J. H. McKay, Midshipman, USN
P. W. Haines, Ensign, USN
S. K. Hall, Ensign, USN
D. Pond, Asst. Surgeon, USNRF
H. R. Delaney, Dental Surgeon, USNRF
A. A. Ammon, Machinist (T), USN
W. R. Roberts, Machinist (T), USN
A. G. Creamer, Carpenter (T), USN
E. C. Grinnell, Midshipman, USN
F. M. Lynch, Midshipman, USN
C. H. Johnson, Midshipman, USN
J. J. Smith, Midshipman, USN
W. A. Lockwood, Midshipman, USN
A. W. Ogle, Captain, USMC
D. Ball, 2nd Lieutenant, USMC
H. B. Emerson, Ensign, USNRF
M. E. Earle, Ensign, USNRF
D. H. Else, Ensign, USNRF
J. S. Fasten, Ensign, USNRF
G. A. Chatel, Ensign, USNRF
L. H. Chase, Ensign, USNRF
C. F. Eddy, Ensign, USNRF
E. F. Ellwood, Ensign, USNRF
E. F. Hoban, Ensign, USNRF
R. N. Russell, Lieut, (jg). Chaplain
W. P. Scott, Captain, USN
F. Y. Weigert, Ensign, USNRF
R. E. Durgin, Ensign, USNRF

U. S. S. Galveston

(Cruiser)

Francis L. Chadwick, Commander, USN
Clarence McC. McGill, Lieut. Comdr., USN
Carl G. Gilliland, Lieutenant, USN
Miitley Perkins, Lieutenant, USN
Wilber M. Lockhart, Lieut, (jg), USN
Owen T. Hurdle, Lieut. (jg) (T) (B), USN
Mertin C. Wade, Jr., Ensign, USN

Brice H. Mack, Ensign (T) (G), USN
Robert J. Ford, Ensign (T) (B), USN
James A. Bass, P. A. Surgeon, USN
Arthur L. Myrland, Asst. Paymaster, USN
Louis M. Bourne, Jr., Captain, USMC
William W. Scott, Captain, USMC
Walter Wilson, Act. Pay Clerk, USN
T. W. Allen, Ensign, USNRF
William F. Andreas, Ensign. USNRF
William L. Hawse, Ensign, USNRF
Joseph B. Hoyt, Ensign, USNRF

Relief Officers in Order of Reporting on Board

T. F. Coleman, Ensign, USNRF
Beverley R. Harrison, Ensign, USNRF
R. W. Collev, Ensign, USNRF
Culber Beebee, Ensign, USNRF
Joseph R. Breasnell, Ensign, USNRF
Waddie P. Jackson, Asst. Surgeon, USNRF
Christopher C. Fewel, Comdr., USN
J. B. McDonald, Jr., Ensign, USNRF
C. McGauly, Ensign, USN
Alfred P. Magness, Lieutenant (jg) (MC), USNRF
Paul F. Christopher, Ensign (T), USN
William C. Eubank, Ensign (T). USN
Kenneth Van House, Ensign (T), USN
Edwin W. Davis, Captain, USCG
Theo. F. Appleby, 2nd Lieut., USMC

U. S. S. George Washington

(Transport)

E. T. Pollock, Captain, USN
G. D. Johnstone, Lieut. Comdr., USN
A. M. Cohen, Lieutenant, USN
J. J. Brosnek, Lieutenant, USN
W. H. Fleming, Lieutenant, USNRF
A. S. Johnston, Lieutenant (jg), USNRF
Morris H. Spriggs, Lieut, (jg), USN
C. E. Milbury, Lieutenant (jg), USNRF
Edward Mansie, Lieut, (jg), USNRF
Clyde Keene, Ensign, USN
J. R. Burkhart. Ensign, USN
Gerald Ollif, Ensign, USN
G. T. Wells, Ensign, USNRF
John F. Murphv, Surgeon, USN
J. G. Enright, P. A. Surgeon, NNV
F. S. Evers, Asst. Surgeon, USN
H. B. Lambert, Asst. Surgeon, NNV

F. G. Pyne, Paymaster, USN
P. B. Wood, Asst. Paymaster, USNRF
J. L. Gillson, Asst. Paymaster, USNRF
R. F. Gates, Asst. Paymaster, USNRF
Donald Hamilton, Asst. Paymaster, USNRF
H. D. Green, Machinist, USNRF
C. E. Rudolph, Pay Clerk, USN
W. D. I. Domer, Asst. Paymaster, USN
R. L. Mabon, Asst. Paymaster, USN
S. C. Girardet, Carpenter, USN
C. E. Rudolph, Pay Clerk, USN
C. A. Wilson, Lieutenant, USNRF
H. F. Tabeling, Lieut, (jg), USNRF
Robert Spearing, Jr., Ensign, USNRF
W. S. Squires, Ensign, USNRF
G. S. Thorp, Pharmacist, USN
J. S. Peters, Lieutenant. USNRF
G. P. Nightingale, Lieut, (jg), USNRF
John R. Roil, Lieutenant (jg), USNRF
J. C. Lough. Dental Surgeon, USN
G. M. Wich, Lieutenant (jg), USNRF
B. B. Brown. Ensign, USNRF
Irving Schwab, Ensign, USNRF
Leon Gottlieb, Ensign, USNRF
B. J. McCann. Ensign, USNRF
C. F. Durgin, Ensign, USNRF
J. F. Massey, Lieutenant (jg), USNRF
Wm. S. Bainbridge, Surgeon, USNRF
Albert Klinger, Lieutenant (jg), USN

Relief Officers in Order of Reporting on Board

A. S. Johnstone, Lieut. Comdr., USNRF
F. B. Stanley, Ensign, USNRF
H. C. Stone, Ensign. USNRF
R. O. Graves, Ensign, USNRF
P. B. Dower. Ensign, USNRF
W. I. Worrell, Lieutenant (jg), USNRF
M. C. Marley, Asst. Surgeon, USN
F. A. Hooker, Asst. Paymaster, USNRF
G. B. Johnson, Dental Surgeon, USNRF I. R.
Meyers, Pay Clerk (T), USN
L. F. Haton, Pay Clerk, USNRF
F. O. Francis, Lieutenant (jg), USNRF
Chas. La Point, Lieut, (jg), USNRF
J. L. Garrity, Ensign, USNRF
R. J. McGinn, Ensign, USNRF
I. Hathaway, Ensign, USNRF
C. L. Winn, Ensign, USNRF
C. R. Denner, Ensign, USNRF
C. C. Gill, Lieut. Commander, USN

Frank Minnikine, Ensign, USNRF
A. C. Adams, Ensign, USNRF
D. S. Boscom, Ensign, USNRF
W. L. Fuller, Ensign, USNRF
A. Barton, Ensign, USNRF
H. O. Roesch, Lieutenant, USN
H. B. Campbell, Ensign, USNRF
J. A. Huston, Ensign, USNRF
L. A. Webber, Ensign, USNRF
F. E. Crawford, Ensign, USNRF
J. Power, Ensign, USNRF
T. B. Van Nest, Ensign, USNRF
W. K. Wortman, Commander, USN
Harmon Hummel, Ensign, USNRF
F. F. Knachel, Ensign, USNRF
J. H. Keith, Ensign, USNRF
C. L. Smith, Ensign, USNRF
R. S. King, Ensign, USNRF
S. Garrison, Ensign, USNRF
C. S. Couchman, Ensign (T), USN
I. Chaplowe, Ensign (T), USN
L. J. Calender, Ensign (T), USN
J. B. Carr, Ensign (T), USN
W. Chandler, Ensign (T), USN
S. O. Cowles, Ensign (T), USN
H. Hodgson, Gunner (E), USN
I. Gray, Lieutenant (jg), USN (MC)
R. P. Dix, Lieut, (jg), USNRF (MC)
E. McCauley, Jr., Captain, USN
F. M. Perkins, Commander, USN
J. R. Shuman, Ensign (T), USN
D. L. Armstrong, Ensign (T), USN
J. F. White, Ensign (PC), USNRF
S. B. Rose, Gunner (E), USN
Alfred Barrett, Machinist, USNRF

G. Grundy, Lieutenant (jg), USNRF
G. Kleinsmith, Ensign (T), USN
C. H. Dougherty, Ensign (T), USN
W. R. McFarland, Ensign (T), USN
H. C. Anderson, Ensign, USNRF
C. H. Porta, Ensign, USNRF
J. B. Wolters, Ensign, USNRF
R. Killman, Ensign, USNRF
R. F. Bradley, Ensign, USNRF
W. A. Dougan, Ensign, USNRF
C. W. Seitz, Pay Clerk (T), USN
H. Mewshaw, Machinist (T), USN
L. B. Provost, Machinist (T), USN
R. P. Helm, Machinist (T), USN
J. Cullinan, Boatswain, USNRF
J. A. Owen, Asst. Surgeon, USN
O. T. Tucker, Pharmacist (T), USN
M. P. Price, Asst. Surgeon, USN
E. D. Richards, Ensign, USNRF
J. P. Jackson, Ensign, USNRF
A. C. Carp, Ensign, USNRF
C. E. Goodhue, Lieutenant, USNRF
E. A. Clair, Ensign, USNRF
W. E. Duncan, Ensign, USNRF
K. K. Weimer, Act." Fay Clerk (T), USN
D. P. Fitzmaurice, Carpenter (T), USN
E. F. Shinn, Act. Pay Clerk (T), USNRF
W. S. Aylesworth, Ensign, USNRF
C. L. Bergstrom, Ensign, USNRF
L. W. Kurtzman, Pharmacist, NNV
J. F. Fairgrave, Ensign, USNRF
J. J. Rafferty, Ensign, USNRF
R. F. H. Crawford, Asst. Paymaster, USNRF
J. W. Parker, Ensign, USNRF
D. G. Smith, Ensign, USNRF

U. S. S. Great Northern
(Transport)

Relief Officers in Order of Reporting on Board

W. W. Phelps, Captain, USN
A. E. Lee, A. P. Surgeon, USN
A. Ahman, Lieut. Commander, USNRF
E. A. Lofquist, Lieutenant, LTSN
D. T. Hunter, Lieutenant, USN
B. E. Tilley, Lieutenant, USN
C. V. McCarty, P. A. Paymaster, USN
C. Wall, Lieutenant, USNRF
W. E. Russell, Lieutenant, USNRF
M. C. Partello, Lieut, (jg), USNRF
J. I. Callanan, Asst. Surgeon, USN
C. R. Barr, Lieutenant, USN
W. A. Anderson, Lieut, (jg), USNRF

J. F. Cox, Lieut. Commander, USN
J. L. Duffy, Lieut. Commander, USN
L. L. Root, Asst. Paymaster, USNRF
F. Black, Ensign, USNRF
W. D. Clarke, Ensign, USNRF
S. Cox, Ensign, USNRF
D. Cummings, Ensign, USNRF
H. H. Dadman, Ensign, USNRF
A. W. De Moulpied, Ensign, USNRF
A. T. Douglas, Ensign, USNRF
W. G. Dow, Ensign, USNRF
V. J. Holsclaw, Machinist, USNRF
P. Paulson, Machinist, USNRF

H. L. Doddin, Midsh. (CBM), USNRF
R. G. Emery, Midsh. (CBM), USNRF
J. A. Hoffman, Midsh. (CBM), USNRF
H. A. Horton, Midsh. (CBM), USNRF
J. Nachman, Midsh. (CBM), USNRF
H. P. Small, Midsh. (CBM), USNRF
W. H. Barowski, Act. Pay Clerk (T), USN
R. H. Roberts, Midshipman, USN
H. A. Rochester, Midshipman, USN
G. C. Whimsett, Act. Chaplain, USNRF
L. Levy, Dental Surgeon, USNRF
F. B. Mullen, Lieutenant (jg), USNRF
C. L. Bergstrom, Ensign, USNRF
H. T. Porter, Machinist, USNRF
A. J. Aver, Machinist, USN
V. J. Holsclaw, Machinist, USNRF
H. B. Buse, Machinist, USNRF
C. K. Hood, Machinist, USNRF
J. S. Cronin, Lieutenant (MC), USN
S. H. R. Dovie, Captain, USN
C. G. Warfield, Lieutenant (PC), USN
Max Cohen, Lieut. (jg) (DC), USNRF
R. W. Clark, Ensign, USNRF
J. G. Coffin, Ensign, USNRF
E. W. Christie, Ensign, USNRF
G. C. Derry, Ensign, USNRF
H. S. Davis, Ensign, USNRF
R. B. Fuller, Ensign (T), USN

U. S. S. Hancock

(Transport)

W. L. Littlefield, Captain, USN
Webb Trammell, Lieutenant, USN
H. D. Hinckley, 1st Lieutenant, USCG
E. M. Williams, Lieutenant, USN
M. G. Gamble, Jr., Lieutenant, USN
G. C. Klein, Lieutenant, USN
Henry B. Broadfoot, Lieutenant, USN
David F. Mead, Lieutenant (jg), USN
James Donaldson, Lieut. (jg), USN
Walter A. Vogelsang, Asst. Surg., USN
Martin J. Costello, Asst. Surg., USN
Lincoln Humphreys, Asst. Surg., USN
Chas. D. Morillon, Pharmacist (T), USN
Harold C. Shaw, P. A. Paymaster, USN
Clifton I. DuFilho, Act. Pay Clerk, USN
Ernest L. Bass, Asst. Nav. Constr., USN
A. J. Rinderle, E. Gunner (T), USN
Chester R. Carson, Gunner, USN
Daniel J. Connor, CM. Gunner, USNRF

Albert F. Weider, Machinist (T), USN Walter
E. Burtt, Lieutenant, USNRF
Robert H. Davison, Ensign, USNRF
Clem S. Clarke, Ensign, USNRF
Louis A. Ferguson, Jr., Ensign, USNRF
Hugh C. Calder, Ensign, USNRF
Guy M. Rothwell, Carpenter, USN
C. E. Strite, P. A. Surgeon. USN

Relief Officers in Order of Reporting on Board

T. C. Smith, Ensign. USNRF
L. D. Causey, Lieut. Commander, USN
L. A. Hopewell, Lieutenant, USNRF
R. M. Anderson, Asst. Paymaster, USN
R. D. Wallerstein. Asst. Paymaster, USNRF
W. R. Ketchum, Ensign, USNRF
E. P. Nickinson, Lieutenant, USN
Wilbert Smith, Commander, USN
J. B. Hardenbergh, Gunner, USNRF
A. T. Moen, Lieutenant, USN
J. E. Erickson, Asst. Paymaster, USNRF
Norman Gaynor, Ensign, USNRF
E. A. Eisele, Ensign (T). USN
E. E. Deane, Ensign (T), USN
J. J. Dall, Ensign (T), I'SN
A. A. Daniels, Ensign (T), USN
E. G. J. Dale, Ensign (T), USN
W. C. Dyer, Ensign (T), USN
Grover C. Klein, Lieutenant, USN

U. S. S. Harrisburg

(Transport)

Wallace Bertholf, Commander, USN
W. W. Turner, Lieut. Commander, USN
Benj. K. Johnson, Lieut. Commander, USN
H. A. Candy, Lieut. Comdr.. USNRF
C. J. Bright, Lieutenant, USN
W. W. Shown, Lieutenant, USN
W. W. Feineman, Lieutenant, USN
E, L. Posey, Lieutenant, USNRF
R. V. Tillett, Lieutenant, USNRF
V. Arntz, Lieutenant, USNRF
W. Joyce, Lieutenant, USNRF
E. M. Post, Lieutenant, USNRF
E. S. Beecher, Lieutenant, USNRF
C. E. Morris, Lieutenant (jg), USNRF
C. F. Smith, Lieutenant (jg). USNRF
John Hynd, Lieutenant (jg), USNRF
John Turner, Lieutenant (jg), USNRF
A. Allen, Lieutenant (jg), USN

C. Keenan, Ensign, USN
W. Graeff, Ensign, USN
T. E. Parsons, Ensign, USNRF
P. J. Guiney, Ensign, USNRF
C. H. Carlson, Ensign, USNRF
H. E. Mayfield, Ensign, USNRF
S. G. Garrett, Ensign, USNRF
W. J. Flower, Ensign, USNRF
H. Van Dyne, Ensign, USNRF
H. A. Lichtenstein, Ensign, USNRF
W. G. Walls, Ensign, USNRF
H. G. Quail, Ensign, USNRF
G. J. Pond, Ensign, USNRF
C. E. Ryder, Med. Inspector, USN
J. G. Powell, Asst. Surgeon, USN
A. O. Sibila, Asst. Surgeon, USN
P. S. McGann, Asst. Dental Surg., USN
B. N. Bradley, Asst. Paymaster, USN
B. M. Cheneweth, Asst. Paymaster, USNRF
H. V. Farnsworth, Asst. Paymaster, USNRF
J. A. Lock, Chief Pay Clerk, USNRF
L. Leonard, Pay Clerk, USNRF
J. J. Lane, Pay Clerk, USNRF
J. A. Anderson, Pay Clerk, USNRF
O. Eng, Boatswain (T), USN
E. Sargent, Boatswain, USNRF
E. W. Sohlman, Gunner (T), USN
C. A. Marlin, Gunner, USNRF
J. A. Kirkpatrick, Pharmacist (T), USN
J. A. Whiteside, Lieut, (jg), USNRF

Relief Officers in Order of Reporting on Board

H. M. Levy, Ensign (T), USN
G. P. McDonald, Ensign (T), USN
R. B. Longyear, Ensign (T), USN
R. C. Lewis, Ensign (T), USN
R. F. McNally, Ensign (T), USN
M. B. Miller, Surgeon, USNRF
R. A. Marshall, Lieutenant, USNRF
D. D. Dewart, Ensign (T), USNRF
R. P. Dodds, Ensign, USNRF
F. B. Snowden, Ensign, USNRF
J. Carroll, Ensign, USNRF
L. T. Forbes, Ensign, USNRF
W. D. Ford, Ensign, USNRF
R. B. Holt, Ensign, USNRF
W. E. Tracy, Ensign, USNRF
W. G. Harrington, Ensign, USNRF
Allen Hetler, Lieutenant (DC), USNRF
W. R. Heymann, Gunner, USNRF
Frank Melson, Gunner, USN

U. S, S. Henderson
(Transport)

G. W. Steele, Jr., Commander. USN
W. C. Barker, Lieut. Commander, USN
F. T. Stevenson, Lieutenant, USNRF
H. D. Dougherty, Lieutenant (jg), USN
Q. R. Thomson, Lieutenant (jg), USN
F. Schultz, Lieutenant (jg), USN
G. Bannerman, Lieutenant (jg), USN
S. Hempstone, P. A. Paymaster, USN
J. L. Barnswell, Lieutenant (jg), USN
S. A. Folsom, Asst. Surgeon, USN
L. S. Hill, Asst. Paymaster, USNRF
E. A. Green, Ensign, USNRF
H. L. Carlson, Ensign, USNRF
C. J. Lamb, Ensign, USNRF
G. F. Pushaw, Ensign, USNRF
G. W. Davis, Act. Pay Clerk, USN
A. Steiner, Carpenter (T), USN
H. A. Horan, Lieutenant, USNRF
E. T. Brown, Lieutenant, USNRF
A. Swendsen, Lieutenant (jg), USNRF
C. J. Hamre, Lieutenant (jg), USNRF
W. D. Sample, Ensign, USNRF
R. McK. Rush, Ensign, USNRF
J. R. McKean, Ensign (T), USN
H. L. Carlsen, Ensign, USNRF
E. A. Green, Ensign, USNRF
G. F. Pushaw, Ensign, USNRF
C. J. Lamb, Ensign, USNRF
N. W. Emery, Ensign, USNRF
J. H. Enwright, Ensign, USNRF
B. J. Eastman, Ensign, USNRF
B. Ellison, Ensign, USNRF
F. H. Flagg, Ensign, USNRF
C. J. Shull, Asst. Paymaster, USNRF
H. S. Goucher, Machinist (T), USN
F. Jeffery, Boatswain, USN
W. H. Frizzle, Gunner (T), USN
W. S. Durkee, Gunner (T), USN
K. D. Legge, Asst. Surgeon, USN
C. T. Frederickson, Asst. Paymaster, USN
W. J. F. Forward, Ensign (T), USN
W. S. H. Hamilton, Ensign (T), USN
A. F. Garrison, Ensign (T), USN
D. T. Duncan, Ensign (T), USN
H. R. Eaton, Ensign (T). USN
C. B. Eddv, Ensign (T), USN
M. C. Doolittle, Ensign (T), USN
W. H. Pate, Boatswain. USNRF

U. S. S. Huntington
(Cruiser)

John K. Robison, Captain, USN
Harry K. Cage, Lieut. Comdr., USN
William D. Greetham, Lieut. Comdr., USN
Robert L. Irvine, Lieut. Comdr., USN
William H. Stiles, Jr., Lieut., USN
Marc A. Mitscher, Lieut, (jg), USN
Reginald S. H. Venable, Lieut. (jg),USN
Henry F. Floyd, Lieut, (jg), USN
Charles E. Rosendahl, Lieut, (jg), USN
Frederick D. Powers, Lieut, (jg), USN
Oliver O. Kessing, Lieut, (jg), USN
Theodore Waldschmidt, Ensign, USN
Hanson E. Ely, Ensign, USN
Andrew I. McKee, Ensign, USN
Lawrence B. Richardson, Ensign, USN
Thomas F. Remington, Ensign, USN
A. Warren Quackenbush, Ensign, USN
Ion Pursell, Ensign, USN
John A. Riley, Ensign (T), USN
Ray Spear, Paymaster, USN
Clarence C. Kress, P. A. Surgeon, USN
Clyde Lovelace, Gunner, USN
George B. Evans, Gunner, USN
Philip A. Astoria, Machinist, USN
Jens Nelson, Machinist, USN
Hector L. Ross, Machinist, USN
Herman R. Newby, Carpenter, USN
Orley Tagland, Pay Clerk, USN
Alonzo H. Woodbine, Comdr., NNV
Warren C. Smith, Lieutenant (jg), NNV
Dwight W. Jennings, Ensign, NNV
Archibald A. Macleish, Surgeon, NNV
George W.Cuthbertson, Lieut., USNRF
Clarence E. Goodhue, Lieut., USNRF
James D. Murray, Lieut., USNRF
James Walsh, Lieut., USNRF
Henry G. Morse, Lieut, (jg), USNRF
Gustayas R. Madden, Lt. (jg), USNRF
Samuel G. Forde, Lieut, (jg). USNRF
William G. Elliott, Ensign, USNRF
Wilfred J. Osborn, Ensign, USNRF
William Halford, Boatswain, USNRF
Albert E. Conner, Machinist, USNRF
Patrick J. Murray, Machinist, USNRF
George W. Eyferth, Pay Clerk, USNRF
Ralph H. Hopgood, Gunner (T), USN
Edward R. McCall, Pharm. (T), USN
Elmer F. Stone, 3rd Lieut. Coast Guard
Robert Donohue, 3rd Lieut. Coast Grd.

Relief Officers in Order of Reporting on Board

Henry W. Hoyt, Lieutenant (jg), USN
John A. Nicol, Carpenter, USN
R. T. Cannon, Asst. Surgeon, USN
R. P. Morse, Dental Surgeon, NNV
Charles C. Copp, Paymaster, USN
William H. Short, P. A. Surgeon, USN
Walter J. Thomas, Ensign (T) (QM), USN
William E. Hooper, Lieut. (E) (jg), USNRF
Walter H. Smith, Lieut. Gg), USNRF
Roy D. Lank, Ensign (E), USNRF
James D. G. Wognum, Act. Pay Clerk, USN
Lester J. Moll, Ensign (T), USN
Foster M. Hampton, Ensign (T), USN
Charles IVL Boswcll, Ensign (T), USN
Samuel M. Haslett, Ensign (T), USN
J. R. Duncan, Lieut, (jg), USNRF
Geo. W. Harrington, Gunner (T) (O),USN
O. T. Mahanay, Gunner (T) (G), USN
W. F. Dickerson, Carpenter (T), USN
J. Burch, Machinist (T), USN
P. J. Chatel. Ensign, USNRF
E. E. Durry, Ensign, USNRF
Herman M. Cohn, Ensign, USNRF
John N. Gatley, Ensign, USNRF
Ferdinand Hunsdorfer, Ensign, USNRF
Chas. S. Leahy, Ensign, USNRF
J. Hunter McDonnell, Ensign, USNRF
Charles N. Steele, Ensign, USNRF
Eugene N. Waeldin, Ensign, USNRF
O. C. Pickney, Ensign, USNRF
Charles I. Stanfield, Ensign, USNRF
William H. Stiles, Lieut. Comdr., USN
J. E. Tracey, Boatswain (T), USN
E. H. Courtney, Ensign. USNRF
Edward S. Kellogg, Captain, USN
Frank McGridge, Lieut, (jg), USNRF
G. W. Burden, Ensign, USNRF
M. Rosenstein, Ensign, USNRF
W. K. Otis, Lieut, (jg) (MC), USNRF
H. O. Johnson, Ensign, USNRF
D. L. Hooker, Ensign, USNRF
H. M. Hood, Ensign, USNRF
R. W. Hunt, Ensign, USNRF
J. B. Harvey, Ensign, USNRF
C. T. Jacobsen, Ensign, USNRF
G. L. Hart, Ensign, USNRF
R. H. Cunningham, Ensign, USNRF
R. C. Deale, Ensign, USNRF

U. S. S. Huron

(Transport)

S. H. R. Doyle, Commander, USN
W. W. Lorshbough, Lieutenant, USN
V. D. Herbster, Lieutenant, USN
A. O. Lustie, Lieutenant, USN
W. Sallsten, Lieutenant (jg), USNRF
C. H. Sargent, Lieutenant (jg), USNRF
W. S. Carrington, Ensign, USN
F. L. Posey, Ensign, USNRF
G. J. Lovett, Ensign, USN
W. J. Carr, Ensign, USNRF
L. Wittman, Ensign, USN
G. L. E. Johnson, Ensign, USNRF
G. E. Richardson, Gunner (E), USNRF
E. A. Zehner, Gunner, USN
J. P. Haynes, Asst. Surgeon, USN
C. L. R. Haines, Asst. Surgeon, USN
D. T. Chalmers, Asst. Surgeon, USN
T. Lyon, Paymaster, USNRF
W. B. Anderson, Asst. Surgeon, USNRF
M. C. Faber, Carpenter, USN
H. E. Parsons, Lieutenant, USNRF
T. H. Pittman, Ensign, USNRF
C. R. Byrne, Machinist, USNRF
W. S. Patterson, Ensign, USNRF
P. J. Dooling, Ensign, USNRF
H. J. Rausom, Pharmacist, USN
R. H. Wakeman, Lieutenant, USN
F. L. Hoffses, Lieutenant (jg), USNRF
E. L. Davezac, Ensign, USNRF
F. F. Tegtmeir, Ensign, USNRF
J. M. Hodder, Ensign, USNRF
D. G. Schmitz, Ensign, USNRF
H. M. Creeger, Carpenter (T), USN
F. L. Worden, Lieutenant, USN
R. L. Armstrong, Ensign, USNRF
E. M. Israel, Ensign, USNRF
E. G. Brain, Asst. Surgeon, USNRF
W. J. Sheridan, Ensign, USNRF
E. J. Rooney, Ensign, USNRF
L. G. Wanderhoof, Ensign, USNRF
G. E. Walthall, Ensign, USNRF
H. E. Ewers, Pay Clerk, USNRF
A. B. Williams, Pay Clerk, USNRF
F. N. Sayre, Ensign, USN
V. H. Schaeffer, Ensign, USN
B. Stebbins, Ensign, USNRF
W. L. Spring, Ensign, USNRF
C. S. Stanley, Ensign, USNRF
C. F. Stephenson, Ensign, USNRF

P. W. Thayer, Ensign, USNRF
G. J. Parker, Asst. Surgeon, USN
L. A. Klauer, Asst. Paymaster, USN

Relief Officers in Order of Reporting on Board

H. E. Jones, Lieut. Commander, USN
F. Johansen, Boatswain, USN
J. P. Hough, Pharmacist, USN
J. M. Bright, Ensign (T), USN
L. F. Edelman, Ensign (T), USN
J. E. Hanahan, Ensign (T), USN
R. E. Farnsworth, Ensign (T), USN
M. F. Ethridge, Ensign (T), USN
J. M. Eisaman, Ensign (T), USN

U. S. S. Isis

(Cruiser Force)
Flagship of Rear-Admiral Marbury Johnston, USN

G. T. Rude, Lieutenant, USNRF
K. E. Nelson, Lieutenant (jg), USNRF
F. I. Peacock, Lieutenant (jg), USNRF
C. K. Green, Ensign, USNRF
F. C. Nyland, Ensign, USNRF
W. C. Wallace, Asst. Paymaster, USN
J. Petterson, Boatswain, USN
P. Kates, Machinist, USN

Relief Officers in Order of Reporting on Board

F. Pardee, Ensign, USN
H. T. Deane, Ensign, USNRF
B. H. Ansell, Ensign, USNRF

U. S. S. Koningen Der Nederlanden

(Transport)

W. M. Hunt, Commander, USN
H. B. Riebe, Lieut. Commander, USN
E. S. McCoach, Lieutenant, USN
W. Mayne, Lieutenant, USN
C. J. Howell, Lieutenant, USN
C. J. Miller, Lieutenant (jg), USN
M. Madsen, Lieutenant (jg), USN
J. C. Ferenz, Lieutenant (jg), USN
W. F. Verleger, Ensign (T), USN
O. H. Maland, Ensign, USNRF

D. Danielson, Ensign, USNRF
J. A. Stevenson, Ensign, USNRF

Relief Officers in Order of Reporting on Board

R. H. Sloan, Lieutenant, USNRF
C. J. Conners, Lieutenant (jg), USNRF
G. Hobbs, Lieutenant (jg), USNRF
R, F. Elder, Ensign, USNRF
H. L. Stevens, Ensign, USNRF
R. Fish, Ensign, USNRF
R. Rollo, Ensign, USNRF
S. C. Gooding, Ensign, USNRF
G. F. O'Hare, Ensign, USNRF
H. H. Vinson, Ensign, USNRF
G. E. Thomas, P. A. Surgeon, USN
H. P. Hare, Asst. Surgeon, USN
G. D. Yoran, Asst. Paymaster, USNRF
H. S. Thayer, Asst. Pavmaster, USNRF
J. Brenner, Gunner (6), USN
H. T. Swimme, Gunner (E), USN
E. F. Smith, Carpenter, USN
J. E. Colthurst, Pay Clerk, USNRF
J. H. Wentworth, Pharmacist, USN
S. Falk, Ensign, USNRF
J. F. Killgrew, Ensign, USNRF
A. F. Alence, Ensign, USNRF
G. S. Wheat, Lieutenant, USNRF
O. M. McNeill. Lieutenant, USNRF
S. B. McDonald, Ensign, USNRF
W. P. Cronan, Captain, USN
H. W. Robertson, Lieut, (jg), USNRF
F. W. Schaill, Ensign, USNRF

U. S. S. Kroonland

(Transport)

M. H. Simons, Commander, USN
H. T. Kays, Lieut. Commander, USN
W. J. Zalesky, Surgeon, USN
E. D, Langworthy, Lieutenant, USN
J. C. Taylor, Asst. Surgeon, USN
A. W. Rutter, Paymaster, USN
R. B. Darling, Lieutenant, USNRF
E. W. Bence, Lieutenant, USNRF
J. R. Lord, Lieutenant, USNRF
L. Serraro, Lieutenant, USNRF
J. G. Hauser, Lieutenant (jg), USN
F. E. Ford, Lieutenant (jg), USNRF
C. H. Bond, Lieutenant (jg), USNRF
M. H. Gill, Lieutenant, (jg) USNRF

M. R. Hudson, Lieutenant (jg), USNRF
A. R. Eubanks, Ensign, USN
R. S. Crocker, Ensign, USNRF
A. Hall, Ensign, USNRF
J. T. MacMurchy, Ensign, USNRF
S. L. Polacheck, Ensign, USNRF
J. L. Garrity, Ensign, USNRF
E. T. Texter, Asst. Surgeon, USNRF
E. B. Welch, Asst. Surgeon, USNRF
C. Wright, Machinist, USN
W. O'Connor, Carpenter, USN
J. R. Tucker. Pay Clerk, LLSNRF
C. R. Shaw, Boatswain, USN
H. Tolderland, Pharmacist, USN
W. E. Hyland, Ensign, USNRF
L. Gluick, Asst. Paymaster, USNRF
C. W. Mannegold, Gunner, USN
J. R. Ostell, Ensign. USNRF
H. D. Hahn, Ensign, USNRF
I. Du Bois, Pay Clerk, USNRF
J. J. S. Fahey, Pay Clerk, USN
W. S. Hyler, Lieutenant, USNRF
F. G. Villmo. Lieutenant (jg), USNRF
J. G. Pettit, Asst. Surgeon, USNRF
C. H. Webber, Dental Surgeon, USNRF
F. M. Coughlan, Ensign, USNRF
J. F. Dixon, Ensign, USNRF
L. A. Rees, Ensign, USNRF
J. F. Nixon, Ensign, USNRF
J. F. Fox, Ensign, USNRF
F. W, Southworth, Asst. Pay., USNRF
H. Nitz, Boatswain, USN.
L. C. Burns, Lieutenant (jg), USNRF
G. L. Buckingham, Lieut, (jg), USNRF
C. Smith, Lieutenant (jg), USN
E. B. Beaumont, Lieut, (jg), USNRF
A. S. Pietre, Ensign, USN
D. A. Powell, Ensign, USN
A. S. Friedman, Ensign, USNRF

Relief Officers in Order of Reporting on Board

R. M. Griswold, Commander, USN
J. Trohear, Lieutenant, USNRF
A. E. Friedman, Ensign, USNRF
H. M. Maxfield, Ensign, USNRF
G. L. Hart, Ensign (T), USN
W. P. Hughes, Ensign (T), USN
R. C. Hunt, Ensign, USN
G. T. Gardner, Ensign, USNRF

W. D. Gallier, Ensign, USNRF
G. Fulton, Jr., Ensign, USNRF
L. P. Gains, Ensign, USNRF
J. H. Donnell, Lieut, (jg), USN (MC)
C. Newman, Lieut. Comdr., USNRF
P. G. Evans, Ensign, USNRF
J. E. Bennett, Ensign, USNRF

U. S. S. Lenape

(Transport)

Robert Morris, Commander, USN
S. A. Manahan, Lieutenant, USN
G. G. Gunderson, Lieutenant, USNRF
E. T. Atkinson, Lieutenant, USNRF
J. W. Phillips, Lieutenant, USNRF
F. A. Mosher, Lieutenant (jg), USNRF
F. R. Myatt, Lieutenant (jg), USNRF
W, A. Soden, Lieutenant (jg), USNRF
N. C. Frey, Lieutenant (jg), USNRF
P. Alvarez, Ensign, USNRF
H. J. Jennings, Ensign, USNRF
O. R. Blair, P. A. Surgeon, NNV
R. E. Watkins, Asst. Surgeon, USN
G. B. Meyers, Asst. Surgeon, USN
E. E. Bell, Asst. Paymaster, USN
J. H. Knapp, Asst. Paymaster, USNRF
A. R. Shearer, Boatswain (T), USN
E. T. Swartz, Gunner (O), USN
W. R. Parish, Gunner (T), USN
A. L. Lund, Machinist (T), USN
G. A. Sipzer, Carpenter, USN
T. A. Morrow, Pharmacist (T), USN
A. G. Shiver, Act. Pay Clerk, USN
W. J. Forrestel, Lieutenant, USN
P. Stinson, Ensign, USNRF
M. Erstad, Ensign, USNRF
J. F. Booth, Ensign, USNRF
J. C. Childs, Ensign, USNRF
S. A. Betts, Ensign, USNRF
L. Chase, Ensign, USNRF
F. Harris, Ensign, USNRF
T. Anderson, Ensign, USNRF
H. Sibley, Ensign, USNRF
O. Christensen, Lieut, (jg), USNRF
W. J. Bowe, Lieutenant (jg), USNRF
J. C. Chambers, Ensign, USNRF

Relief Officers in Order of Reporting on Board

B. G. Barthalow, Commander, USN
H. H. Fowler, Lieutenant (jg), USN

E. L. Glozier, Ensign, USNRF
O. H. Ehrmann, Ensign, USNRF
F. D. Devlin, Lieutenant (jg), NNV
J. L. Collins, Ensign, USNRF
L. W. Wilber. Ensign, USNRF
H. W. Leshin, Asst. Surgeon, USN
J. W. Kelliher, Ensign, USN
A. H. Gilbert, Ensign, USN
E. T. King, Ensign, USN
R. L. Gill, Ensign, USN
H. Bradley, Ensign, USN

U. S. S. Leviathan

(Transport)

J. W. Oman, Captain, USN
W. N. Jeffers, Commander, USN
J. H. Blackburn, Lieut. Comdr., USN
Edwin Altheiser, Lieut. (jg), USNRF
E. W. Andrews, Ensign (t), USN
R. J. Ast, Asst. Paymaster, USNRF
Edward J. Amberg, Asst. Paymaster, USNRF
H. A. Cunningham, Lieutenant, USNRF
E. J. Carroll, Asst. Surgeon, USN
D. Coughlan, Boatswain, USNRF
T. S. Coulbourn, Asst. Paymaster, USN
H. Davidson, Lieutenant, USNRF
A. B. Dorsey, Ensign (T), USN
H. J. Edwards, Ensign (T), USN
J. Foster, Lieutenant, USNRF
J. W. Ford, Lieutenant, USNRF
A. J. Gahagan, Ensign (T), USN
W. L. Graeff. Ensign (T), USN
O. L. Hankinson, Lieutenant, NNV
F. K. Harper, Lieutenant (jg), USNRF
A. E. Harding, Lieutenant (jg), USNRF
L. G. Hoffman, Asst. Paymaster, USN
E. M. Hudson, Asst. Surgeon, USN
O. J. W. Haltnorth, Ensign (T), USN
F. Hannon, Machinist, USN
H. B. Judkins, Asst. Paymaster, USNRF
J. H. Jack, Carpenter, USN
E. D. Jones, Ensign, USNRF
John Jones, Lieutenant, USNRF
C. A. Krez, Ensign, USN
George Keeser, Ensign (T), USN
W. Lau, Ensign (T). USN
F. D. Manock, Lieutenant, USN
1. Nordstrom, Ensign (T), USN
J. A. O'Donnell, Gunner, USNRF
J. C. Parker, Ensign (T), USN

J. J. Snyder, Surgeon, USN
G. C. Schafer, Paymaster, USN
W. H. F. Schluter, Ensign (T), USN
W. E. Smith, Pay Clerk, USNRF
C. W. Smith, Boatswain, USNRF
C. R. Shannon, Gunner, USNRF
W. J. Thomas, Asst. Paymaster, USNRF
Geo. T. Vaughan, Surgeon, USNRF
V. V. Woodward, Lieutenant, USN
Jas. Watson, Lieutenant, USNRF
F. S. Watt, Lieutenant (jg), USNRF
C. W. Waters, Act. Pay Clerk, USN
C. H. Boucher, Lieutenant, USN
A. H. Bateman, Ensign, USN
J. L. Beebe, Ensign, USNRF
R. C. Bright. Ensign (T), USN
W. A. Barber, Asst. Paymaster, USNRF
W. M. Benton, Pharmacist (T), USN
LeRoy B. Foster, Asst. Paym., USNRF
P. C. Grening, Lieut. Comdr.. USNRF
Edwin F. Barker, Asst. Paym., USN
E. J. Alexander, Asst. Paym., USN
V. J. Gunnell, Asst. Paymaster, USN
C. W. Waters, Asst. Pay Clerk, USN
M. Bergman, Gunner (T), USN
E. D. Heinz, Gunner (T), USN
Robt. Martin, Pharmacist (T), USN
F. L. Rector, Boatswain (T), USN
Cunningham, Lieutenant, USNRF
W. Davidson, Lieutenant, USNRF
J. Foster, Lieutenant, USNRF
J. W. Ford, Lieutenant, USNRF
C. Hembey, Lieutenant, USNRF
A. E. Alexander, Lieut, (jg), USNRF
R. G. Skead, Lieutenant (jg), USNRF
Geo. V. Tawes, Lieutenant (jg), USNRF
C. E. Cadmus, Ensign, USNRF
O. E. Cobb, Ensign, USNRF
De Coursey Fales, Ensign, NNV
C. S. Ziesel, Dental Surgeon, USN
W. L. F. Simonpietri, Paymaster, USN
L. D. Miller, Lieutenant (jg), USNRF
S. B. Nicholas, Ensign, USNRF
A. J. Sherlock, Ensign, NNV

Relief Officers in Order of Reporting on Board

A. B. Fry, Captain. USNRF
A. F. Foss, Lieutenant (jg), USNRF
G. F. Poggi, Pay Clerk, USNRF
H. F. Bryan, Captain, USN

Richard H. Jones, Lieutenant, USN
R. J. Lorentz, Asst. Surgeon, USN
A. K. Dunlap, Asst. Surgeon, USN
W. C. Looncy, Lieutenant (jg), USNRF
C. E. Cadmus, Ensign, USNRF
William Seward Allen, Ensign, USNRF
S. V. B. Nichols, Ensign, USNRF
J. R. Ditmars, Ensign, USNRF
E. C. Ferguson, Ensign, USNRF
P. F. Howe, Ensign, USNRF
H. A. Mann, Ensign, USNRF
M. R. Falin, Ensign, USNRF
R. C. Seaman, Ensign, USNRF
E. H. Thompson, Ensign, USNRF
J. W. Shuler, Asst. Paymaster, USNRF
L. B. O'Shaughnessy, Asst. Paymaster, US-NRF
G. F. Poggi, Pay Clerk, USNRF
E. E. McDonald, Chaplain, USN
H. A. May, Surgeon, USN
N. B. Farwell, Paymaster, USN
C. I. Campbell, Chief Pharmacist, NNV
S. L. Froehlich, Ensign, USNRF
Nelson Gay, Ensign, USNRF
T. A. Gaynor, Ensign, USNRF
IT. B. Judkins, Asst. Paymaster, USNRF
H. M. Turner, Lieutenant (ig), USNRF
R. H. Knight, Ensign, USNRF
F. D. K. LeCIerq, Ensign, USNRF
C. B. Carlon, Ensign, USNRF
W. J. Armiger, Ensign, USNRF
H, R. Ingram, Ensign, USNRF
J. W. Barcus, Ensign, USNRF
T. S. Schad, Ensign, USNRF
R. F. Beardsley, Ensign, USNRF
L. F. Singleton, Ensign, USNRF
C. C. Cox, Ensign, USNRF
E. H. Durell, Ensign, USN
A. T. Leonard, Lieutenant, USN
L. J. Arnold, Ensign (T), USN
L. J. Ewbank, Ensign (T), USN
G. R. Fitzsimmons, Ensign (T), USN
E. S. Croasdale. Ensign (T), USN
J. M. Ferry, Ensign (T), USN
C. W. Waters, Ensign, USN
J. H. Willey, Ensign, USNRF
T. C. Memington, Lieut, (jg), USNRF (PC)
W. G. Burgess, Lieut, (jg), USN (PC)
D. A. Huges, Ensign, USN
G. W. Brown, Ensign, USN
R. W. Bockius, Ensign, USN
T. B. Morehouse, Ensign, USN

V. Barringer, Lieutenant (T), USN
W. Moore, Lieutenant (jg) (T), USN
J. Nelson, Lieutenant, USN
J. K. Billingsley, As.st. Paym., USNRF
J. M. Baker, Asst. Paymaster, USNRF
J. W. Shuler, Asst. Paymaster, USNRF
E. M. Crofutt, Asst. Surgeon, USNRF
W. E. Malloy, Lieutenant, USN
W. A. Dundon, Machinist, USN
F. G. Wright, Lieutenant. USNRF
A. C. Fapkin, Ensign, USNRF
S. Morrill, Ensign, USNRF
F. J. Stephens, Asst. Paym., USNRF
E. J. Wriglev, Asst. Paymaster, USNRF
W. W. Phelps, Captain, USN
R. B. Haines, Ensign, USN
J. E. Porter, Lieutenant, USN
J. G. Deacon, Ensign, USNRF
A. Grant, Ensign, USNRF
J. C. Evans, Ensign, USNRF
M. L. Lequin, Ensign, USNRF
D. F. Milan, Ensign, USNRF
C. H. Miller, Ensign (PC), USNRF
C. K. Osborne, Lieutenant, USN
C. M. Hammond, Ensign, USN
L, F. Leventhal, Ensign, USN
H. L. Howell, Lieutenant, USN
C. A. Soars, Lieutenant, USN
C. C. Hilliard, Lieutenant (jg), USNRF
A. Braunwarth, Boatswain, USNRF
J. F. Maegher, Ensign, USNRF
A. F. Vare, Ensign, USNRF
G. G. Johnston, Ensign, USNRF

U. S. S. Louisville
(Transport)

J. P. Jackson, Commander, USN
R. M. Comfoit, Lieut. Comdr., USN
H. Hartley, Lieutenant, USN
G. C. Cummings, Lieutenant, USN
C. L. Meek, Lieutenant, USNRF
W. H. Missett, Lieutenant, USNRF
G. E. Stay, Lieutenant, USNRF
J. Carstairs, Lieutenant, USNRF
F. W. Sievers, Lieutenant, USNRF
E. E. Merrill, Lieutenant, USNRF
R. H. Dumphe, Lieutenant (jg), USNRF
A. W. Walls, Lieutenant (jg), USNRF
A. Malcolm, Lieutenant (jg), USNRF
E. H. Bruns, Lieutenant (jg), USNRF
A. P. Canning, Lieut, (jg). USNRF

A. Greiner, Lieutenant (jg), USNRF
J. Stirling, Lieutenant (jg), USNRF
R. Strassburger, Lieut, (jg), USNRF
C. C. Kimball, Surgeon, NNV
J. V. McAlpin, Dental Surgeon, USN
C. W. Eley, Asst. Dental Surgeon, USN
J. R. Byrne, Asst. Dental Surg., USN
R. B. Scribner, Asst. Paymaster, USN
S. Trimble, Asst. Paymaster, USNRF
L. B. Beatty, Asst, Paymaster, USNRF
P. C. Harvey, Ensign, USNRF
G. W. Hutchins, Ensign, USNRF
G. D. Taylor, Ensign, USNRF
W. R. Lyon, Ensign. USNRF
F. L. Wooley, Ensign, USNRF
W. J. Keenan, Ensign, USNRF
H. G. Frank, Ensign, USNRF
E. H. Sanborn, Ensign, USNRF
S. C. Beckwith, Ensign, USNRF
G. Berton, Ensign (T), USN
B. S. Riley, Ensign, USN
G. A. Cartwell, Ensign, USNRF
J. Elsman, Ensign, USNRF
F. B. Wise, Ensign, USNRF
S. L. Thomas, Ensign, USNRF
W. P. McCartv, Ensign, USNRF
W. M. Shaughnessy, Ensign, USNRF
O. D. Forawalt, Pharm. (T), USN
T. T. Taylor, Asst. Pay Clerk, USN
W. B. Hinckley, Asst. Pav Clerk, USN
G. P. North, Gunner, USNRF
R. J. Collins, Gunner (T), USN
F. J. Nofs, Gunner (T), USN
W. V. Tynan, Carpenter (T), USN
F. J. Woods, Carpenter (T), USNRF
H. J. Osman, Machinist, USNRF
R. J. Proudfoot, Machinist, USNRF

Relief Officers in Order of Reporting on Board

L. F. Carleton, Lieutenant (jg), USNRF
G. F. Taylor, Ensign, USNRF
P. M. Childs, Ensign, USNRF
R. N. Munley, Ensign, NNV
A. W. Lockwood, Ensign, USNRF
M. B. Antrim, Ensign, USNRF
R. H. Becker, Ensign, USNRF
P. M. Chase, Ensign, USNRF
F. C. Wildman, Ensign, USNRF
A. O. Rule, Ensign, USN
J. W. Roper, Ensign, USNRF

C. B. Bare. Act. Chaplain. USN
F. X. Lynch, Act. Pay Clerk, USN
J. E. Goodwin, Lieut, (jg), USNRF
A. B. Rivers, Dental Surgeon, USNRF
M. Clements, P. A. Surgeon, USNRF
J. T. McBreen, Ensign, USNRF
L. F. Maughan, Ensign, USNRF
C. C. Roehan, Ensign, USNRF
R. W. Miller, Ensign, USNRF
G. T. Murphy, Ensign, USNRF
F. J. Seefurth, Ensign, USNRF
E. L. Lawlor, Ensign, USNRF
W. H. Bennitt, Ensign, USNRF
F. W. Culver, Ensign, USNRF
W. B. Hinkley, Act. Pay Clerk, USNRF
H. Hathaway, Ensign, USNRF
J. K. Haviland, Ensign, USN
G. J. Hawk, Ensign, USN
J. R. Lopez, Ensign, USNRF
A. Larch, Ensign, USNRF

U. S. S. Madawaska

(Transport)

E. H. Watson, Commander, USN
C. McCauley, Lieutenant, USN
W. B. Cothran, Lieutenant, USN
J. M. Shoemaker, Lieut, (jg), USNRF
A. W. Bird, Ensign, USN
F. G. Abeken, Surgeon, USN
T. Williamson, Jr.. P. A. Paym., USN
F. M, Smith, Asst. Nav. Constr., USN
E. L. Jones, Boatswain, USN
D. McCallum, Gunner, USN
S. H. Sacker, Ensign, USN
P. H. Cassidy, Machinist, USN
R. McD. Moser, Lieutenant, USNRF
M. A. McPhee, Lieut, (jg), USNRF
A. E. Dunham, Ensign, USNRF
J. O. Dunham, Ensign, USNRF
T. A. Jeffrey, Ensign, USNRF
L. W. Shaffer. Asst. Surgeon, USNRF
R.W. Auerbach, Asst. Surgeon, USNRF
S. E. Hall, Ensign, USNRF
J. C. Work, Ensign, USNRF
F. W. Stoker. Ensign, USNRF
Edw. C. Bliss, Lieutenant, USNRF
William Gorman, Ensign, USNRF
E. F. Blain, Carpenter, USNRF
H. D. Meeker, Surgeon. USNRF
S. B. Flvnn. Asst. Paymaster, USNRF
F. S. Sullivan, Pay Clerk, USN

I. B. Talton, Asst. Paymaster, USN
H. M, Mason, Asst. Paymaster, LTSN
R. P. Huntington. Lieut, (jg), USNRF
R. A. Brett, Lieutenant (jg), USNRF
M. J. Carton, Carpenter, USN
L. D. Crandon, Asst. Paym., USNRF
W. Armour, Ensign, USNRF
J. K. Carr, Ensign, USNRF
A. M. Henshaw, Lieutenant, USNRF
R. T. Luce, Lieutenant, USNRF
C. M. Schwab. Ensign, USNRF
F. T. McCarthy, Ensign, USNRF
F. W. Bloscher, Ensign, USNRF
A. F. O'Brien, Ensign, USNRF
A. J. Gallagher, Ensign, USNRF
Wm. S. Thomas, P. A. Surgeon, USNRF
J. J. Murphv, Ensign, USNRF
D. M. Herbert, Ensign, USNRF

Relief Officers in Order of Reporting on Board

E. T. Constien, Commander, USN
L. Foust, Ensign, USNRF
E. A. Stevens, P. A. Surg. Asst., USN
T. Nielson, Lieutenant (jg). USNRF
I. E. Pitman, Ensign, USNRF
W. N. Thomas, Acting Chaplain, USN
W. A. Bailey, Pay Clerk, USN
W. S. Brady, Asst. Surgeon, USN
L. G. Szarmanaki, Boatswain, USN
Charles McKenna, Pay Clerk. USNRF
A. W. Brunner, Ensign, USNRF
C. M. Smith, Ensign, USNRF
F. F. Thacher, Ensign, USNRF
A. B. Hayward, Surgeon, USN
E. Lanois, Asst. Surgeon, USN
B. M. Hendrickson, Gunner, USN
H. J. Mullenhagen, Asst. Surg., USN
Robert Henderson, Commander, USN
Dwight Tenney, Lieut, (jg), USNRF
C. H. McDonald, Gunner (T), USN
A. W. Hinckley, Ensign, USNRF
R. A. Newman, Ensign (T), USN
R. A. Light, Ensign (T), USN
A. W. Liddle, Ensign (T), USN
B. Lewis, Ensign (T), USN
A. C. Headley, Ensign (T), USN
K. J. Blundon, Ensign (PC), USNRF
H. B. Lee, Ensign (T), USN
O. Brooks, Lieutenant (MC), USN
M. Olcott, Ensign (PC). USNRF
R. F. Luce, Lieut. Comdr., USNRF

U. S. S. Mallory

(Transport)

H. Williams, Commander, USN
H. E. Knauss, Lieut. Commander, USN
R. E. Keating, Lieutenant, USN R. L. Low,
Lieutenant, NNV A. G. Burt, Lieutenant, US-
NRF W. H. Scollan. Lieutenant, USNRF W. W.
Verner, P. A. Surgeon, USNRF H. B. La Favre,
P. A. Surgeon, USN
D. R. Davidson, Asst. Surgeon, USN M. J. Di-
erlam, Lieut, (jg). USNRF
E. R. Glosten, Lieut, (jg), USNRF A. Burr,
Lieut, (jg), USNRF
W. D. I. Domer, Asst. Paymaster, USN W. J.
Bork, Ensign, USNRF
E. W. Bowne, Ensign, USNRF
C. A. Crane, Ensign, USNRF
J. A. McCarthy, Ensign, USNRF
E. Fife, Ensign, USNRF
M. J. Peralta, Ensign, USNRF
J. A. McBridge, Boatswain, USNRF
G. T. Thornton, Gunner, USN
A. F. Threm, Gunner, USN
I. Streger, Machinist, USN
C. B. Porter, Machinist, USNRF
L. B. Karelle, Pay Clerk, USN
E. E. Nelson, Carpenter, USN
A. E. Chase, Asst. Paymaster, USNRF
W. O. Sprout, Pharmacist, USN
E. H. Von Heimburg, Ensign, USN

Relief Officers in Order of Reporting on Board

W. C. Vose, Ensign, USN
R. R. Greenwood, Ensign, USNRF
G. S. Harrison, Ensign, USNRF
P. A. Harrison, Ensign, USNRF
A. C. Handy, Ensign, USNRF
P. E. Mackett, Ensign, USNRF
N. D. Godfrey, Ensign. USNRF
M. L. Hard, Ensign, USNRF
C. K. Reinke, Asst. Surgeon, USN
F. L. Stiles, Lieutenant (jg), USNRF
M. Blaken, Lieutenant (jg), USNRF
A. Burr, Lieutenant (jg), USNRF
R. Haguewood, Pharmacist, USN
R. L. Holland, Lieutenant (jg), USNRF
H. H. Rairden, Ensign (T), USN
E. W, Lotz, Lieutenant (jg), USNRF
E. J. Kinkard, Lieut, (jg) (MC), USN
J. D. McCrea, Ensign, USNRF

H. I. MacKon, Ensign, USNRF
A. E. Raynor, Ensign, USNRF
G. E. McCall, Ensign, USNRF
B. Lubic, Ensign, USNRF
J. C. Lott, Ensign, USNRF

U. S. S. Manchuria

(Transport)

C. S. Freeman, Commander, USN
W. W. Waddell, Lieut. Comdr., USN
A. Zeeder, Lieut. Comdr., USNRF
J. W. Rakow, Lieutenant, USNRF
S. H. Hurt, Lieutenant (jg), USN
F. C. Neal, Lieutenant (jg), USNRF
W. A. Thom, Lieutenant (jg), USNRF
E. Standish, Lieutenant (jg), USNRF
A. Nobel, Lieutenant (jg), USNRF
L. A. Van Matre, Lieut, (jg), USNRF R.
Rowles, Lieutenant (jg), USNRF
E. S. Clark, Ensign, USNRF V. P. Suttelle,
Ensign, USNRF
F. J. Baumgartner, Ensign, USNRF
C. B. Hannum, Ensign, USNRF
R. G. Buschatzky, Ensign, USNRF
J. S. Woodward, Medical Insp., USN
L. W. Shaffer, Asst. Surgeon, USN
E. J. Alexander, Asst. Paymaster, USN
J. F. Van Duren, Asst. Paymaster, USN
R. E. Rockett, Machinist (T), USN
C. E. Keptner, Gunner (Ord.), USN
C. F. Hudson, Gunner (EL), USN
W. Bittner, Boatswain (T), USN
H. McAlmond, Carpenter, USN
G. D. Sipe, Pharmacist, USN
W. F. Shaw, Act. Pay Clerk, USN

Relief Officers in Order of Reporting on Board

J. V. Lynn, Asst. Surgeon, USNRF
D. F. Condrick. Asst. Paym. USNRF
L. B. Cranz, Dental Surgeon, USNRF
G. W. Gaffney, Ensign, USNRF
E. W. Johnston, Ensign, USNRF
R. C. Outten, Pay Clerk, USNRF
L. A. Straits, Ensign, USNRF
H. C. Reed, Ensign, USNRF
R. C. Hay, Ensign, USNRF
A. C. Haven, Ensign, USNRF
W. C. Heppenheimer, Ensign, USNRF
D. M. Hill, Ensign, USNRF

A. Hickey, Ensign, USNRF
A. J. Mesmer, Ensign, USNRF
H. M. Carey, P. A. Surgeon, USNRF
F. L. Sample, Carpenter (T), USN
F. H. Ogle, Pharmacist (T), USN
H.M. Scull, Ensign, USN
M. T. Seligman, Ensign, USN
C. G. Quillian, Lieutenant, USNRF
M. Greenleaf, Lieutenant, USN
G. M. Dennis, Ensign, USNRF
A. C. Stevens, Ensign, USNRF
G. W. Shepard, Surgeon, USN
A. G. Ruff. Ensign, USN
J. A. Robinson, Ensign, USN
E. T. McHenry, Ensign, USN
W. H. Mackay, Ensign, USN
R. J. Mackay, Ensign, USN
H. R. Mack, Ensign, USN
C. O. Johnson, Machinist, USNRF
K. G. Shiels, Machinist, USNRF
W. P. Richardson, Machinist, USNRF
H. P. Krummes, Asst. Surgeon, USN

U. S. S. Martha Washington
(Transport)

Chauncey Shackford, Comdr., USN
Schamyl Cochran, Lieutenant, USN
Charles S. Root, Lieutenant, USCG
Francis H. Hardy, Lieutenant, USNRF
Benjamin C. Judd, Lieutenant, USNRF
Oscar J. \Mieeler, Lieutenant, USNRF
Amos B. Root, Lieutenant (jg), USN
Frank G. Fahrion, Lieutenant (jg), USN
Robert E. Wilkinson, Asst. Nav. Constr. (T), USN
George A. Berry, Lieut, (jg), USNRF
Fred W. Davis, Lieut, (jg), USNRF
Samuel N. Sinclair, Lieut, (jg), USNRF
Leonard Roll, Ensign (T), USN
Alexander Stuart, Ensign (T), USN
George W. Waldo, Ensign (T), USN
Andrew Skinner, Ensign (T), USN
Henry E. Rung, Ensign (T), USN
Ten Eyck H. Reed, Ensign, USNRF
Hobart P. Swanton, Ensign, NNV
Robert B. Bruce, Ensign, USNRF
Daniel L. Chamberlain, Ensign, USNRF
Wilbur J. Clark, Ensign, USNRF
Maurice M. Bennett, Ensign, USNRF
Montgomery A. Stuart, P.A. Surg. USN
William F. McAnally, Asst. Surg., USN

Harbeck Halsted, Asst. Surg., NNV
Andrew Mowat, Asst. Paym., USN
John A. Joseph, P. A. Paym., NNV
Donald P. Smith, Asst. Paym., USNRF
Hubert G. Webb, Asst. Paym., NNV
F. W. Larkworthy, Machinist (T), USN
Charles H. Burch. Machinist, USNRF
Alexander E. Meigs, Machinist, USNRF
Armand Mayville, Carp. (T). USN
William N. Landrum, Pharm. (T) USN
Chauncey J. Buckley, Act. Pay Clerk (T), USN
Howard J. Hoffman, Pay Clerk, USNRF

Relief Officers in Order of Reporting on Board

Frank Kinne, Ensign (T), USN
Joseph Hall, Boatswain (T), USN
Edward T. Comins, Lieut. Qg), USN
Lloyd P. Burgess, Lieut, (jg), USNRF
Vincent P. O'Connell, Ensign, USNRF
Arthur W. Rand, Ensign, USNRF
Kenneth M. Rendall, Ensign, USNRF
Arthur H. Patten, Pay Clerk, USNRF
Isadora C. Woodward, Act. Chapl.,USN
Isador J. F. Dubois, Chief Pay Clerk, USNRF
Philip E. Wait, Ensign, USNRF
Howard L. Tibbetts, Ensign, USNRF
Eugene C. Tirrell, Ensign. USNRF
Charles R. Vinton, Ensign, USNRF
Frank V. Uhrig, Ensign, USNRF
Joseph L. Walsh, Ensign. USNRF
C. J. Rend, Midshipman, USN
T. G. W. Settle, Midshipman, USN
Ellery R. Fitch, Lieut, (jg). USNRF
D. H. Casto, Surgeon, USN
James E. Britt, Asst. Paym., USNRF
Orlando D. Reed, Boatswain (T), USN
Paul W. Georges, Boatswain (T), USN
Arthur E. Lawrence, Carp, (T), USN
Stephen M. Henagan, Mach. (T), USN
Coleman Marshman, Gunner (T), USN
T. H. Sharp, Asst. Surgeon, USN
Martin P. Kane, Lieut, (jg) (DC), USNRF
K. G. Castleman, Captain, USN
Samuel J. Mealy, Ensign (T), USN
Charles W. Mario w. Ensign (T), USN
Edmund B. Montgomery, Ensign (T), USN
Whitney W. Miller, Ensign (T), USN
L. A. Rice, Ensign (T), USN
Allyn W. Maxwell, Ensign (T), USN
Henry Sosvielle, Boatswain (T), USN

U. S. S. Matsonia

(Transport)

J. M. Luby, Captain, USN
E. H, Williams, Lieut. Comdr.. USN
R. McD. Moser, Lieut. Comdr.,USNRF
E. R. Henning, Lieutenant, USN
N. Fogarty, Lieutenant, USNRF
C. E. Tabrett. Lieutenant, USNRF
P. E. Kuter, Lieut, (jg), USN
W. E. McClintock, Lieut, (jg), USN
W. L. Fawcett, Lieut, (jg), USNRF
N. T. Short, Lieut, (jg), USNRF
T. E. Bray, Lieut, (jg), USNRF
J. B. Cadenbach, Ensign, USN
A. J. Baiter, Ensign, USNRF
E. O. Blomquist, Ensign, USNRF
I. W. Murray, Ensign, USNRF
C. E. Stone, Ensign, USNRF
G. Dowdle, Ensign, USNRF
A. F. Westman, Ensign, USNRF
R. P. Bentley, Ensign, USNRF
R. D. Bickford, Ensign, USNRF
F. C. Beck. Asst. Paymaster, USN
A. P. Tibbetts. P. A. Surgeon, NNV
H. W. Harris, Asst. Surgeon, USN
J. E. Malcolmson, Asst. Surgeon, USN
P. H. Levey, Asst. Paymaster, USNRF
F. Anderson, Asst. Paymaster. USNRF
S. M. Thompson, Gunner (T), USN
J. J. Brierly, Gunner. (T) (E) USN
E. G. Williams, Carpenter (T), USN
P. Le Van, Machinist (T), USN
R. M. Dumphy, Pharmacist, USN

Relief Officers in Order of Reporting on Board

H. T. Daniels, Lieutenant (E), USNRF
W. J. Wheatley, Lieut, (jg), USNRF
E. C. Hartup, Pay Clerk, USN
F. D. Armstrong, Ensign, USNRF
L D. Nagel. Ensign, USNRF
P. Streeter, Ensign. USNRF
J. E. Sharpe, Ensign, USNRF
J. E. Malcolmson, Asst. Surgeon, USN
H. M. Cone, Asst. Paymaster, USNRF
C. E. Short, Ensign, USN
D. J. Sinnott, Ensign, USN
S. W. Higgins, Ensign, USNRF
R. E. Holden, Ensign, USNRF
G. E. Hodge, Ensign, USNRF

H. K. Leventon, Lieutenant, USN
F. Schweitzer, Act. Chaplain, USN
M. L. Frizelle, Ensign (Eng.), USNRF
C. G. Reeves. Ensign (Eng.), USNRF
M. P. Kane, Dental Surgeon, USNRF
J. Jelke, Jr., Asst. Paymaster, USNRF
L. Smith, Boatswain (T), USN
J. P. Jackson, Commander, USN
A. R. Murray, Ensign, USNRF
J. T. Davies, Ensign, USNRF
R. G. Starck, Ensign, USNRF
C. W. Scranton, Ensign, USNRF
C. B. Fenton, Ensign, USNRF
J. W. Sullivan, Ensign, USNRF
C. C. Stoeber, Ensign (Eng.), USNRF
J. F. Yoes, Act. Pay Clerk (T), USN
W. Rind, Lieut. Commander, USNRF
R. S. Parr, Lieutenant, USN
C. P. Moriartv, Ensign (T), USN
W. S. Moore, Ensign (T), USN
A. C. Moysey, Ensign (T), USN
J. A. Ryan, Lieut, (jg) (Eng.), USNRF
J. Silverman, Ensign (T) (Eng.), USN
C. A. Schellens, Ensign (T) (Eng.), USN
J. A. Regnier, Lieut, (jg) (Dent. Corps), USNRF
John Sweetland, Ensign, USNRF
A. H. Flickwir, Lieut, (jg) (Med. Corps) USNRF

U. S. S. Maui

(Transport)

C. A. Abele, Commander, USN
E. B. Woodworth, Lieut. Comdr., USN
W. F. M. Edwards, Lt. Comr., USNRF
H. F. Councill, Lieutenant, USN
J. P. Rasmussen, Lieutenant, USNRF
A. Ryan, Lieutenant, USNRF
V. H. WTieeler, Asst. Paymaster, USN
R. M. Little, P. A. Surgeon, USNRF
P. Keller, Asst. Surgeon, USN
R. W. Lewis, Asst. Surgeon, USNRF
A. H. Westerberg, Lieut, (jg), USNRF
R. W. Dunham, Lieut, (jg), USNRF
P. R. Griffin, Asst. Paymaster, USNRF
J. L. Kershaw, Ensign, USN
A. Squires, Ensign, USNRF
J. T. Viegas, Ensign, USNRF
S. H. Robinson, Ensign, USNRF
E. C. Reed, Ensign, USNRF
D. Weir, Ensign, USNRF

J. Marmion, Ensign, USNRF
J. W. R. Stewart, Ensign, USNRF
E. L. La Dieu, Machinist, USNRF
W. J. McFate, Ensign, NNV
P. D. Boore, Gunner, USN
A. McGraw, Gunner, USN
G. B. Martinsen, Bosn., USN
J. J. Maune, Carpenter, USN
F. B. Bork, Pharmacist, USN
W. S. Rockwell, Act. Pay Clerk, USN

Relief Officers in Order of Reporting on Board

C. K. Patterson, Ensign, USNRF
C.B. Gosnell, Asst. Paymaster, USNRF
E. R. Carrol, 1st Lieut., USAQMC
E. H. Sandelin, Lieutenant, USNRF
T. P. Wynkoop, Ensign, USN
R.E. Jones, Ensign, USNRF
J. M. Keep, Ensign, USNRF
V.R. Hood, Ensign, USNRF
C. E. Howland, Ensign, USNRF
J.T. Keegan, Ensign, USNRF
E. S. Huntley, Ensign, USNRF
H. S. Woodman, Ensign, USN
J.T. Low, Asst. Surgeon, USNRF
C. S. Freeman, Captain, USN
E. J. O'Toole, Ensign (T), USN
J. A. Pentz, Ensign (T), USN
T. A. O'Connor, Ensign (T), USN
W. F. J. Odenwald, Ensign (T), USN
E. F. Thrall, Ensign (T), USN
F. W. Stirzel, Ensign (T), USN
W. J. O'Hara, Ensign (T), USN
Leo A. Redmond, Ensign (T), USN
P. C. Hulse, Dental Surgeon, USNRF
Walter E. Hennerich, Asst. Surg., USN
Randle Clifford, Asst. Surgeon, USN
W. S. Rockwell, Act. P. C. (T), USN

U. S. S. Mercury

(Formerly Barbarossa) (Transport)

H. L. Brinser, Commander, USN
P. P. Bassett, Lieut. Comdr., USN
H. A. Arnold, Lieutenant, USNRF
E. A. McIntyre, Lieutenant, USN
T. F. Webb, Lieutenant (jg), USNRF
H. F. Kent, Lieutenant (jg), USNRF
D. C. Woodward, Ensign, USN
D. Duncan, Ensign, USN

P. R. Fox, Ensign, USN
P. E. Kuter, Ensign, USN
E. J. Frieh, Ensign. USN
E. M. Shipley, Ensign, USN
G. R. Crapo, Paymaster, USN
C. W. O. Bunker, P.A. Surgeon, USN
R. J. Bower, Asst. Surgeon, USN
R. E. Watkins, Asst. Surgeon, USN
K. E. F. Sorenson, Machinist, USN
J. Holler, Machinist, USNRF
W. A. Nightingale, Carpenter, USN
C. P. Hines, Pharmacist, USN
E. Dann, Chief Pay Clerk, USN

Relief Officers in Order of Reporting on Board

D. Heath, Lieutenant (jg). USNRF
E. N. Dwight, Ensign, USNRF
William James, Ensign, USN
W. Prior, Asst. Paymaster, USN
A. H. White, Asst. Paymaster, USN
W. Johnson, Machinist (T), USN
H. S. Lyons, Lieutenant, USNRF
J. C. Warrington, Lieut, (jg), USN
B. R. LeRoy, Ensign, USNRF
A. D. Freshman, Ensign (G), USN
F. S. Jameson, Asst. Surgeon, USNRF
E. S. Walker, Lieutenant (jg), USNRF
G. E. Atkinson, Ensign (D), USNRF
T. J. Turney, Ensign (E), USNRF
H. A. Arnold, Lieutenant, USNRF
E. A. Salisbury, Lieutenant, USNRF
H. L. Look, Lieutenant, USNRF
L. M. Willson, Lieut, (jg), USNRF
R. O. Brackett, Lieut, (jg), USNRF
S. F. French, Lieut, (jg), USNRF
E. Carroll, Lieut, (jg), USNRF
J. C. Waage, Jr., Ensign, USNRF
G. A. Schaub, Asst. Surgeon, USNRF
P. J. Doyle, Asst. Paymaster, USNRF
J. S. Spaven, Ensign, LTSN
A. T. Sprague, Ensign, USN
S. D. Starbuck, Ensign, USNRF
B. S. Wilson, Ensign, USNRF
L. B. Wheeler, Ensign, USNRF
A. K. Wardwell, Ensign, USNRF
E. F. Woodward, Ensign, USNRF
F. M. Bansom, Asst. Paym., USNRF
E. A. Salisbury, Lieutenant, USNRF
H. C. Berkstresser, Carpenter, USN
W. D. Brereton, Lieut. Comdr., USN
G. T. Rude, Lieutenant, USNRF

T. J. Turney, Lieut, (jg), USNRF
R. C. Midwood, Dental Surg., USNRF
O. S. Powell, Asst. Paymaster, USN
D. W. Mitchell, Asst. Paymaster, USN
W. F. Robins, Pay Clerk, USNRF
D. J. Reilly, Boatswain (T), USN
E. R. Murphy, Elec. Gunner (T), USN
H. M. Cohn, Cadet
R. S. Lloyd, Cadet A. M. Turney, Cadet
M. F. Hersion, Cadet
A. E. Chatterton, Cadet
James Sterling, Lieut, (jg), USNRF
J. A. Sweeney, Ensign, USNRF
H. C. Miller, Ensign (PC), USN
F. F. Babcock, Machinist (T), USN
W. A. Anderson. Lieut, (jg), USNRF
F. A. Mullen, Ensign (T), USN
H. W. Neely, Ensign (T), USN
H. V. Nussey, Ensign (T), USN
H. R. Spofford, Ensign, USNRF
B. E. Belcher, Lieut. (MC), USN
E. J. Lanois, Lieutenant (MC), USN
A. W. Hagman, Pay Clerk. USNRF

U. S. S. Minneapolis
(Cruiser)

H. H. Christy, Capta,in, USN
Gordon W. Haines, Lieut. Comdr., USN
George H. Bowdey, Lieutenant, USN
Harry A. Badt, Lieutenant, USN
Cleaveland C. Kimball, Surg., USNRF
Roger D. DeWolf, Lieutenant, USNRF
Edgar F. Marbourg, Asst. Paym., USN
Proctor M. Thornton, Ensign, USN
Ralph S. Riggs, Ensign, USN
Lisle Henifin, Ensign, USN
James Morrison, Ensign, USN
James R. Selfridge, Ensign, USNRF
Joseph R. Williams, Ensign, USNRF
Augustus W. Walker, Ensign, USNRF
Albert W. Hinckley, Ensign, USN
Charles King, Boatswain, USN
Curry Eason, Gunner, USN
John Gallagher, Machinist, USN
Edwin H. Briggs, Machinist, USN
Ernest F. Kiefer, Carpenter, USN
Houston S. Stubbs, Pay Clerk, USN
Clifton I. DuFilho. Act. Pay Clerk, USN

Relief Officers in Order of Reporting on Board

Thomas Shine, Lieutenant (jg), USN
Charles F. Manley, Carpenter, USN
J. F. Hines, Commander, USN
Aner Erickson, Act. Pay Clerk(T), USN
Marion W. Jones, Act. Carp. (T), USN
Walker W. Anderson, Ensign, USNRF
Harold S. Johnson, Ensign, USNRF
Elwood O. Langill, Ensign, USNRF
Joseph B. Lindquist, Ensign, USNRF
R. Z. Johnston, Captain, USN
L. C. Covell, 1st Lieutenant, USCG
William B. Dortsch, Lieut, (jg), USNRF
H. V. Bressler, Lieut, (jg), USNRF
F. M. Orton, Ensign, USNRF
J. B. Clapp, Ensign, USNRF
L. S. Taylor, Ensign, USNRF
J. S. Bauman, Ensign, USNRF
D. R. Hudson, Ensign, USNRF
H. A. Young, Ensign, USNRF
A. C. Smith, Ensign, USNRF
W. M. Allen, Ensign, USNRF
I. M. Jacobs, Asst. Surgeon, USN
Edward P. Grisbacker. Mach. (T), USN
Lloyd T. Chalker, Captain, USCG
Robert T. Young, Lieutenant, USN
Wm. J. Johnston, Lieut, (jg), USNRF
Ralph W. Hungerford, Ensign, USN
James J. Hughes, Ensign, USN
Albert P. Beals, Ensign, USNRF
Kenneth H. Bayliss, Ensign, USNRF
Joseph E. Derosier, Ensign, USNRF
Byron M. Fleming, Ensign, USNRF
William E. Dodge, Ensign, USNRF
Luther H. Elliott, Ensign, USNRF
Frank Durand, Jr., Ensign, USNRF
James P. Harland, Ensign, USNRF
Thomas Eraser, Ensign, USNRF
Claude J. Geisel, Ensign, USNRF
Albert O. Mang, Machinist (T), USN
Floyd J. Sexton, 1st Lieut., USCG
Mattheas A. Roggenkamp, Asst. Pay Clerk (T). USN
Charles Blanchard, Boatswain, USNRF
Lewis H. C. Johnson, Ensign, USNRF
Harrison A. Jones, Ensign, USNRF
Adolph J. Woll Webber, Gunner (R), USN
Hugh M. Kitchen, Ensign, USNRF
David W. Jones, Ensign, USNRF
N. S. Knight, Ensign. USNRF

F. E. Kennedy. Ensign. USNRF
H. W. Kephart, Ensign, USNRF
C. E. Chillingworth, Ensign, USNRF
Charles P. Snyder, Captain, USN
C. D. Holland, Lieut, (jg), USN
Alexander Steel, Asst. Paym., USNRF
Joseph A. Farrell, Act. Pay Clerk, USN
John H. Herke (PC), USN
Hobert Hankinson, Ensign, USNRF
Gilbert L. Duchars, Ensign, USNRF
J. E. Robertson, Ensign, USNRF

U. S. S. Mongolia

(Transport)

Willis McDowell, Commander, USN
H. McL. Walker, Lieut. Comdr., USN
Emery Rice, Lieut. Comdr., USNRF
J. D. Smith, Lieutenant, USN
V. J. Green, Lieutenant, USNRF
W. Tornroth, Lieutenant, USNRF
P. W. Bond, Lieutenant, USNRF
B. Christenson, Lieutenant, USNRF
J. G. Lutz, Lieutenant (jg), USNRF
S. S. Green, Lieutenant (jg), USNRF
W. B. Anderson, Ensign, USN
P. J. Dooling, Jr., Ensign, USNRF
E. W. Higgins, Ensign, USNRF
R. M. Treco, Ei.sign, USNRF
S. N. Danskin, Ensign, USNRF
P. D. Oulton, Ensign, USNRF
A. A. Lofquist, Ensign, USNRF
E. L. Blake. Ensign, USNRF
H. T. Bryant, Gtmncr, USN
H. Holcombe, Carpenter, USNRF
F. Joseph, Boatswain, USNRF
J. A. Materlick, Lieutenant, USNRF
J. W. Merget, Lieutenant (jg), USN
A. Treux, Lieutenant (jg), USNRF
J. F. Hatton, Lieutenant (jg), USNRF
W. C. Thierbach, Ensign (E), USNRF
E. E. Powers, Ensign (E), USNRF
A. J. Iverson, Ensign (E), USNRF
J. J. Carroll, Ensign (E), USNRF
D. H. Noble, P. A. Surgeon, USN
H. P. Griffin, Asst. Surgeon, USN
H. H. Slominski, Asst. Surgeon, USN
A. R. Leh, Pharmacist, USN
H. A. Daniels, Dentist Surgeon, USN
L. H. Huebner, P.A. Paymaster, USN
T. B. Mudd, Asst. Paymaster, USNRF
H. R. Tiffany, Asst. Paymaster, USNRF

E. S. Gilbert. Pay Clerk, USN
E. J. Horn. Pay Clerk, USNRF

*Relief Officers in Order of Reporting on
Board*

C. K. Wildman, Ensign, USN
F. O. Willenbucher, Ensign, USN
D, G. Methany, Asst. Surgeon, USNRF
F. B. Goddard, Lieut, (jg), USNRF
Charles P. Snyder, Commander, USN
A. W. Lindstrom, Gunner, USN
Gideon J. Ellis, Lieut, (jg), USNRF
Leslie M. Shorter, Midsh., USNRF
R. L. Smith, Midshipman, USNRF
R. L. Van Siclen, Midshipman, USNRF
R. F. Wilson,_ Midshipman, USNRF
L. Henefin, Lieutenant, USN
J. K. Batchelder, Ensign, USNRF
F. E. Covalt, Lieutenant (jg), USNRF
A.W. Yowell, Ensign (Pay Clk.) USNRF
H. B. Porterfield, Ensign, USN
C. E. Peterson, Ensign, USN
C. W. Proctor, Ensign, USN (T)
E. C. Peterson, Ensign, USN (T)
R. H. Persons, Ensign, USN (T)
W. H. Stueve, Ensign, USN (T)
E. C. Steinhart, Lieut, (jg). USNRF
Frank Barth, Ensign, USNRF

U. S. S. Montana

(Cruiser)

Chester Wells, Commander, USN
R. E. Pope, Commander, USN
A. W. Sears, Lieut. Commander, LSN
A. J. James, Lieut. Commander, USN
H. P. Glover, Lieutenant, USN
S. M. Kraus, Lieutenant, USN
H. M. Branham, Lieutenant (jg), USN
H. H. Bouson, Lieutenant (jg), USN
M. Case, Lieutenant (jg), USN
J. D. Edwards, Ensign, USN
W. N._ McDowell, P. A. Surgeon, USN
E. Stein, Asst. Surgeon, USN
W. S. Zane, P. A. Paymaster, USN
H. Schmidt, Captain, USMC
J. Evans, Boatswain, USN
L. Rodd, Chief Gunner, USN
C. B. Bradley. Gunner, USN
G. Bradley, Gunner, USN

L. O. Peterson, Gunner, USN
W. H. Langdon, Machinist, USN
H. H. Fowler, Machinist, USN
J. Chinnis, Machinist, USN
W. Collins, Carpenter, USN
W. Craig, Chief Pay Clerk, USN
S. B. Deal, Act. Pay Clerk, USN
J. M. Damrow, Act. Pay Clerk, USN
L. A. Puckett, Act. Pay Clerk, USN
J. J. Miffitt, Act. Pay Clerk, USN
A. G. Robinson, Lieutenant (jg), USN
H. R. Gellerstedt, Lieut, (jg), USN
E. S. McCawley, Lieutenant (jg), USN
L. D. Pickering, Lieutenant (jg), USN
W. S. Hactor, Ensign, USN
F. L. Lowe, Ensign, USN
T. L. Nash, Ensign, USN
A. R. Earley, Ensign, USN
P. W. Yeatman, Ensign, USN
R. S. Wyman, Lieutenant (jg), USN
R. de S. Horn, Ensign, USN
K. Preston, Ensign, USN
H. G. Eldredge, Ensign, USN
F. B. Smith, Ensign, USN
T. E. Chandler, Ensign, USN
H. V. Wiley, Ensign, USN
T. G. Peyton, Ensign, USN
C. W. Flynn, Ensign, USN
A. C. Thomas, Ensign, USN
R. M. Fortson, Ensign, USN
L. Wood, Ensign, USN
S. A. Maher, Ensign, USN
J. M. Field, Ensign, USN
J. H. Sprague, Ensign, USN
H. S. Clark, Ensign, USN
A. Landis, Ensign, USN
M. O. Carlson, Ensign, USN
J. W. Simms, Ensign, USN
C. O. Bain, Gunner, USN
E. F. Gumm, Gunner, USN

Relief Officers in Order of Reporting on Board

W. R. Nichols, Ensign, USN
V. R. Murphy, Ensign, USN
G. D. Chester, Ensign, USN
S. C. Norton, Ensign, USN
E. H. Krueger, Ensign, USN
H. V. Deely, 2nd Lieutenant, USMC
L. L. Babbitt, Lieutenant, USN
M. A. Deans, Ensign, USN
W. J. Skelton, Ensign, USNRF

W. A. Stillwell, Ensign, USNRF
E. PL Gibbs, Asst. Surgeon, USNRF
S. D. Moyer, Carpenter, USNRF
H. B. Cecil, Lieutenant, USNRF
P. J. A. Leduc, Acting Chapl., USNRF
C. R. Brown, Gunner (T), USN
B. J. Reynolds, Gunner (T), USN
J. O. Crom, Machinist (T), USN
M. R. Hinkle, Machinist (T), USN
A. C. Perring, Carpenter (T), USN
J. M. Damrow, Act. Pay Clerk, USN
J. L. Foss, Act. Pay Clerk (T), USN
W. Liggett, Jr., Lieut. Comdr. (Ret.), USN
R. J. Miller, Lieutenant (ig), USN
W. E. Findeisen, P. A. Surgeon, USN
J. A. Mangiaracina, Asst. Surg., USN
H. B. Ransdell, P. A. Paymaster, USN
D. McK. Paulson, Ensign, USNRF
T. P. Wells, Ensign, USNRF
J. L. D. Painter, Ensign, USNRF
T. F. Kilkenny, Jr., Ensign, USNRF
G. T. Morrow, Ensign, USNRF
W. G. Woodams, Ensign, USNRF
A. M. Baldwin, Ensign, NNV
G. I. Wright, Lieut, (jg), USNRF
L. H. Van Syckle. Ensign, NNV
E. F. Cloney, Ensign, NNV
C. R. Jacobson, Ensign, NNV
A. P. Flagg, Lieutenant (jg), USN
F. C. Cobb, Lieutenant (jg), USNRF
A. P. Hill, Lieutenant (jg), USNRF
C. S. Shields, Ensign, USNRF
C. A. Painter, Ensign, USNRF
E. J. McKiernan, Machinist (T), USN
S. C. Norton, Lieutenant (jg), USN
D. Ballard, Ensign (T), USN
J. F. Hart, P. A. Surgeon, USN
K. Hamner, Dental Surgeon, USNRF
C. R. Sies, Asst. Paymaster, USN
T. L. Zynda, Gunner (T). USN
W. F. Rail. Gunner (T), USN
H. J. Megin. Pharmacist (T). USN
L. Ashcroft, Ensign. USNRF
J. B. Clark, Ensign, USNRF
E. L. Denton, Ensign, USNRF
P. P. Kane, Ensign, USNRF
L. W. Fisher, Ensign, USNRF
M. E. Miller, Ensign. USNRF
I. C. Mix, Ensign, USNRF
L. B. Roberts, Ensign, USNRF
F. C. Smith, Ensign, USNRF
D. Sparkman, Ensign, USNRF

190

S. B. Sharp, Machinist (T), USN
J. J. Graham, Ensign, USN
F. H. Gilmer, Ensign, USN
W. E. Moser, Boatswain (T), USN
H. F. A. Baske, Asst. Surgeon, USN
R. G. Brown, Ensign, USNRF
A. S. Francis, Ensign, USNRF
A. E. Spinner, Gunner (T), USN
T. J. Hassett, Gunner (T), USN
G. C. Day, Captain. USN
H. E. Kays, Commander, USN
J. S. Spore, Lieut. Commander, USN
J. T. Lett, Ensign. USNRF
J. B. McGovern, Ensign, USNRF
E. B. Luckie, Ensign. USNRF
H. A. Lincoln, Ensign. USNRF
R. L. Gray, Ensign, USNRF
H. P. Dockstader, Ensign, USNRF
H. R. Lake, Ensign, USNRF
G. C. Lindberg, Ensign, USNRF
H. C. Cooper, Captain, USMC
C. A. McGaha, Boatswain (T), USN
H. W. Bryan, Boatswain (T), USN
R. E. Woods, Gunner (T), USN
A. H. Giesler, Carpenter (T), USN
A. W. Robbins, Pay Clerk, USNRF
L. B. Roberts, Ensign, USNRF
C. F. Tinney, Lieutenant (M), USNRF
C. M. McAfee, Ensign, USNRF
L. A, Krake, Ensign, USNRF

U. S. S. Mount Vernon

(Transport)

A. N. Robertson, Captain, USN
R. Morris, Commander, USN
J. M. Doyle, Lieutenant, USN
P. L. Carroll, Lieutenant (jg), USN
W. W. Feineman, Ensign, USN
R. G. Heiner, Surgeon, USN
J. P. Kutz, Paymaster, USN
D. n. Sumner, Lieut. Comdr.. USNRF
F. C. Bailey, Lieutenant. USNRF
P. A. Guttormsen, Lieutenant, USNRF
T. L, McAvery, Lieutenant, USNRF
G. Tyner, Lieutenant, USNRF
J. n. Gorman, Lieutenant (jg), USNRF
E. Rock, Lieutenant (jg), USNRF
S. L. Almon, Ensign. USN
J. A. Martin, Ensign (T), USN
J. C. Stein, Ensign (T), USN
R. G. Baird, Ensign, NNV

G. W. Milliken, Ensign, USNRF
R. de B. Clark, P. S. Surgeon, NNV
W. M. Anderson, Asst. Surg., USNRF
R. W. Harris, Asst, Paymaster, USNRF
V. B. Havens, Asst. Paymaster, USNRF
H. V. C. Wetmore, Asst. Nav. Constr. (T), USN
F. C. A. Plagemann, Bosn., USN
R. L. Marshall. Bosn., USNRF
C. H. Kohls, Gunner, USN
R. J. Youngkin, Pharmacist (T), USN
M. J. Dambacher, Pay Clerk, USN
H. Noble, Pay Clerk, USNRF

Relief Officers in Order of Reporting on Board

R. W. Emmons, Lieutenant, USNRF
R. Roberts, Lieutenant, USNRF
L. Heyl, Lieutenant (jg), USNRF
W. J. Pittuck, Lieutenant (jg). USNRF
P. C. Taisey, Ensign. NNV
H. O. K. Hansen, Ensign, NNV
V. H. Robinson. Ensign, USNRF
A. E. Rowe, Bosn.. USNRF
J. T. Ogden. Gunner, USNRF
V. F. Le Verne, Ensign (T) (M), USN
A. Blake. Ensign, USNRF
C. K. Cummings, Ensign, USNRF
J. R. Hooper, Jr., Ensign, USNRF
G. W. Eastman, Asst. Surgeon, USNRF
J. G. Prout, Asst. Paymaster, USN
H. M. Shaffer, Asst. Paymaster, USN
Lon H. Robb, Bosn. (T), USN
J. W. Rabbitt, Gunner (T), USN
M. Witte, Gunner (T), USN
A. F. Sortwell, Bosn., USNRF
T. A. Clark, Carpenter (T), USN
H. A. D. Cameron. Lieutenant, USNRF
E. C. W. S. Lyders, Lieutenant, USNRF
W. Comerford. Lieut, (jg), USNRF
A. T. Appleyard, Ensign, USNRF
R. H. Baker. Ensign, USNRF
W. L. Freeborn, Ensign, USNRF
E. M. Wallman, Ensign, USNRF
W. Wilson, Ensign, USNRF
G. A. Smith, Dental Surgeon, USNRF
A. Mangin, Warrant Officer (Fr. Navy)
D. E. Dismukes, Captain, USN
J. P. Friefer, Ensign, USNRF
A. N. Hanau, Ensign, USNRF
J. S. Hanna, Ensign, USNRF
H. L. Morrison, Ensign, USNRF
A. Staton, Lieut. Commander, USN

F. Myers, Lieuteaant (jg), USN
F. J. McCarthy, Act. Surgeon, USN
C. E. Brown, Act. Surgeon, USN
B, T. Smith, Asst. Paymaster, USNRF
W. A. Hopkins, Act. Chaplain, USN
R. J. Pose, Carpenter, USN
H. H. Hersh, Machinist. USNRF
D. C. Moore, Pharmacist, USNRF
J. A. Campbell, Ensign, USNRF
B. V. McGovern, Ensign, USNRF
D. W. Dodd, Asst. Surgeon, USNRF
S. Weinstein, Gunner (T), USN
J. F. Nelligan, Ensign, USNRF
K. M. Fiske, Ensign, USNRF
H. H. McNeill, Ensign, USNRF
W. P. Ives, Ensign, USNRF
A. R. Tierney, Ensign, USNRF
P. E. Covalt, Ensign, USNRF
H. H. McGlaughlin, Ensign, USNRF
S. D. Barr, Act. Pay Clerk, USN
M. H. Stein, Ensign, USN
A. R._ Staudt, Ensign, USN
J. Briggs, Ensign, USN
H. D. Knower, Ensign, USNRF
L. D. Crouter, Asst. Paymaster, USNRF
W. V. C. Brandt, Ensign, USNRF
W. S. Porteous, Jr., Ensign, USNRF
L. Placet, Boatswain, USN
E. E. Curtis. Surgeon, USN
T. W. Anthony, Ensign, USNRF
R. B. Zinser, Ensign, USNRF
P. P. Dreffein, Ensign, USNRF
J. J. Ward, Ensign, USNRF
D. F. Gang, Ensign, USNRF
R. I. Law, Pharmacist, USNRF
E. W. Bentley, Ensign (PC), USNRF
R. B. Scharmon, Ensign, USNRF
M. A. Malandian, Lieut, (jg), USNRF

U. S. S. New Orleans
(Cruiser)

Arthur G. Kavanagh, Commander, USN
Harold A. Waddington, Lieut., USN
Daniel J. Callaghan, Lieut., USN
John L. Riheldaffer. Lieut., USN
Oscar W. Leidel, P.A. Paymaster, USN
Robin B. Daughtry, Lieut, (jg), USN
Walker Cochran, Lieut, (jg), USN
Ross T. McIntire, Asst. Surgeon, USN
Edward A. Mitchell, Ensign, USN
Byron K. Presnell, Ensign, USN

Leonard B. Austin, Ensign, USN
Stanley M. Haight, Ensign, USN
Walter L. Taylor, Ensign, USN
Oscar L. Youngblood, Pay Clerk, USN

Relief Officers in Order of Reporting on Board

Waldo Evans, Captain, USN
Kenneth M. Bennett, Commander, USN
Charles Higginson, Ensign, USN
George D. Howell, Ensign, USN
Bryant H. Howard, Ensign, USN
Ralph B. Bristol, Asst. Paym., USN
W. J. Bisel. Pharmacist, USN
Lewis P. Scott. Ensign (T), USN
Morris Phinney, Ensign (T), USN
Pursel J. Earl, Act. Pay Clerk, USN
Alfred S. Reynolds, Ensign, NNV
Howard G. Muzzy, Ensign, USNRF
Bruno F. Miller, Ensign, USNRF
Joseph L. Regnier, Dent. Surg., USNRF
Edgar B. Larimer, Commander, USN
A. P. Moran, Jr., Ensign, USN
V. R. Moore, Ensign, USN
D. G. Calif, Ensign, USNRF
Robert H. Whitaker, Act. Pay Clerk (T), USN
Milton N. Price, Ensign, USNRF
Robert O. Hinckley, Lieut, (jg), USN
Charles R. Jennings, Gunner, USNRF
Charles J. McLean, Machinist, USNRF
Wilbur M. McKay, Ensign (T), USN
James McKillips, Ensign (T), USN
Harlan H. Grover, Ensign (T), USN
Philip D. Werum, Asst. Surg. (Lt.), USN
Charles H. Mecum, Lieutenant, USN
Frank G. Eldridge, Machinist (T), USN

U. S. S. North Carolina
(Cruiser)

W. T. Tarrant, Commander, USN
H. F. Emerson, Lieutenant, USN
R. F. Frellsen, Lieutenant, USN
R. E. Cassidy, Lieutenant, USN
R. S. Geiger, Captain, USMC
D. A. McElduff, Lieutenant, USN R. E.
Sampson, Lieutenant, USN
W. M. Corry, Lieutenant (jg), USN
H. T. Bartlett, Lieutenant (jg), USN
G. L. Woodruff, Lieutenant (jg). USN
R. Asserson, Ensign, USN
R. Burhen. Ensign, USN

E. F. McCartin, Ensign, USN
J. J. Clark. Ensign, USN
E. A. Foote, Ensign, USN
F. E. Duyall, Jr.. Ensign, USN
J. H. Duncan, Ensign. USN
E. G. Higgins, Ensign, USN
J. W. Merget. Ensign, USN
R. H. Bush, Ensign, USN
W. A. Sherman, Lieutenant (jg), NNV
W. A. Anderson, Lieut, (jg), NMC
G. Norman, Ensign, USNRF
A. J. R. Ferguson, Ensign, USNRF
C. J. Ingersoll, Ensign, USNRF
C. S. Bartow, Ensign, USNRF
P. S. Dennis, Ensign, USNRF
O. E. Cobb, Ensign, USNRF
H. A. Courtney, Ensign, USNRF
D. D. Cooke, Ensign, USNRF
M. J. Looram, Ensign, USNRF
J. Hallas, Ensign, NNV
W. J. Charles, Ensign, NNV
L. Doane, Ensign. NNV
H. C. Marshall, Ensign, NNV
R. S. Bunker, Ensign, NNV
H. W. Smith. P.A. Surgeon, USN
C. L. McCarthy, Asst. Surgeon, USN
J. L. Chatterton, P.A. Paym., USN
W. R. Thomas, Asst. Nav. Constr.,USN
F. Rasmussen, Boatswain, USN
R. B. England, Gunner (E), USN
A. K. Goffe, Gunner (0), USN
F. P. Kenny, Pay Clerk, USN

Relief Officers in Order of Reporting on Board

F. E. O'Brien, Asst. Surgeon, USN
C. G. A. Halwartz, Boatswain, USN
C. F. Holzermer, Gunner (Ord.) (T),
USN F. S. Miller, Gunner (Radio) (T), USN
A. W. Chandler, Machinist (T), USN
T. M. Waldschmidt, Lieut, (jg), USN
R. N. Hedges, Asst. Surgeon, USN
E. W. Amos, Carpenter (T), USN
R. F. Jones. P. A. Surgeon, USN
E. F. Humphrey, Ensign, USNRF
W. R. Thompson, Ensign, USNRF
H. E. Ulich, Ensign, USNRF
T. H. Banks, Ensign, USNRF
E. M. Burstan, Captain, USMC
A. W. Harrington, Captain, USMC
H. N. Heine, Lieutenant (jg), USNRF
M. G. Wolson, Lieutenant (jg), USNRF

A. C. Brattle, Lieutenant (jg), USNRF
E. Priodrick, Lieut. Commander, USN
C. L. Turner, Lieutenant (jg), USNRF
W. D. MacDougall, Captain, USN
F. W. Eiker, 2nd Lieutenant, USMC
R. B. Lanier, Ensign, USNRF
J. B. Kingsley, Ensign, USNRF
G. R. Osborne, Ensign, USNRF
W. H. Holby, Ensign, USNRF
D. F. Parker, Ensign, USNRF
C. E. Seage, Lieutenant (jg), NNV
J. Hallas, Ensign, NNV
R. S. Bunker, Ensign, NNV
G. K. Mesick, Ensign, USNRF
J. J. Roth. Ensign, USNRF
S. M. Lang, Ensign, USNRF
R. W. Belknap, A. Surgeon, USN
B. K. Johnson, Lieut, Comdr., USN
M. J. Peterson, Lieut. Comdr., USN
W. E. Brown, Lieut. Comdr., USN
G. A. Hasler, Lieutenant, USN
E. M. Parker, Ensign, USNRF
R. F. Wliitlock, Ensign, USNRF
H. O. Baker, Ensign, USNRF
J. D. Matthews, Ensign, USNRF
A. Szarmanski, Boatswain, USN
R. A. Whitney. Ensign, USNRF
G. D. Wheeler, Ensign, USNRF
R. H. Schooley, Ensign, USNRF
A. F. Soukup, Ensign, USNRF
C. H. Westaby. Ensign, USNRF
A. McK. Willson, Ensign, USNRF
T. J. Farley, Ensign, USNRF
E. H. Pauson, Pay Clerk, USNRF
H. L. Reed, Ensign, USNRF
W. G. Evans, Ensign, USNRF
B. B. Williams, Ensign, USNRF
R. E, Leaver, Gunner. USN

U. S. S. Northern Pacific

(Transport)

C. F. Preston, Captain, USN
A. T. Hunter, Lieut. Comdr., USNRF
B. B. Taylor, Lieut. Comdr., USN
G. J. Lenhardt, Lieutenant, USNRF
R. S. Smith, Lieutenant (jg), USNRF
J. F. Peters, Ensign, USNRF
IL V. Van Dusen, Ensign, USNRF
W. I. Green, Ensign, USNRF
Walter Clayton, Ensign, USNRF
Otis Wood, Ensign, USNRF

R. G. Davis, P. A. Surgeon, USN
John I. Ballinger, Ensign (M), USN
W. H. A. Pike, Lieutenant, USN
Thomas M. Cassidy, Ensign (T), USN
August Anderson, Ensign (T), USN
A. D. Denney, Lieut., USN, Gun. Off.
R. E. Rogers, Lieut., USN, Engr. Off.
W. E. Fitzgerald, Lieut., (jg), USN. (T) (C)
H. T. Mulloy, Machinist, USN
I. V. Herin, Ensign (T) (G), USN
G. S. Bull, Ensign, USNRF
R. G. Morsell, P.A. Paymaster, USN
W. L. Bunker, Lieutenant, USNRF
D. R. Haguewood, Pharmacist, USN
R. S. Mecklem, Asst. Paym., USNRF

Relief Officers in Order of Reporting on Board

A. S. Freedman, Asst. Paymaster USN
T. F. Crockett, Boatswain, USNRF
W. Richardson, Asst. Surgeon, USNRF
Charles Hierdahl, Ensign (G), USN
I. W. Thompson, Pay Clerk, (T) USN
M. J. Price, Asst. Surgeon, USN (Lt.)
J. A. B. Sinclair, Act. Asst. Surg., USN
G. V. Clark, Asst. Surgeon, USNRF
J. B. Morris, Lieutenant, USNRF
John C. Ruddock, Asst. Surgeon, USN
J. K. Davis, Lieutenant, USN
Reuben Jasperson, Pay Clerk, USNRF
F. Seefeldt, Ensign, USN
H. V. Van Dusen, Lieut, (jg), USNRF
A. C. Bristol, Ensign (D), USNRF
F. H. Robb, Ensign (D), USNRF
D. S. Martinez, Ensign (D), USNRF
R. H. Wilcox, Ensign (D), USNRF
F. D. Allen, Carpenter, USN
E. V. Byrne, Chaplain, USN
L. O. Colbert, Lieutenant, USNRF
F. G. Hogan, Asst. Paymaster, USNRF
T. M. Cassidy, Lieut, (jg), USN
M. P. Kane, Dental Surgeon, USNRF
H. Busching, Machinist, USNRF
A. Lindstrom, Machinist, USNRF
A. G. Latimer, Ensign, USNRF
H. P. Knickerboker, Ensign, USNRF
P. R. Ladd, Ensign, USNRF
A. E. Koch, Ensign, USNRF
W. R. Read. Midshipman, USN
W. H. Mays, Midshipman, USN
E. J. Martin, Lieut, (jg), USNRF
C. L. Howard, Lieut, (jg), USNRF

J. J. Calkins, Lieut, (jg). USNRF
A. W. Mackenzie, Ensign, USNRF
F. J. Massey, Ensign, USNRF
F. W. Sartain, Ensign, USNRF
C. Morse, Ensign, USNRF
S. S. Fosnaugh, Bosn. (T), USN
H. M. Norton, Gunner (T), USN
A. J. Hockman, Machinist (T), USN
A. H. Richter, A. P. Clerk (T), USN
H. D. Burroughs, A. P. Clerk (T), USN
R. D. McNeill, Bosn. (P), USNRF
C. E. Morgan, Lieutenant, USNRF
J. G. Das, Asst. Surgeon, USNRF
W. T. Davidson, Asst. Surgeon, USN
O. J. Hackler, Gunner (T), USN
G. E. Fithen, Ensign, USNRF
G. E. Gilmour, Ensign, USNRF
A. E. McMahon, Ensign, USNRF
D. R. Lowry, Ensign, USNRF
W. F. Higgins, Ensign, USNRF
F. M. Hill, Ensign, USNRF
A. S. Herbert, Engn, USNRF
W. J. Kistle, Machinist, USNRF
S. P. Folsom, Machinist, USNRF
D. L. Gill, Ensign, USNRF
B. P. Huske, Lieut. (Chapl.), USNRF
P. J. Lynch, Carpenter, USN
E. M. Foote, P. A. Surgeon, USN
T. J. Boner, Pharmacist, USN

U. S. S. Olympia

(Cruiser)

Bion B. Bierer, Captain, USN
George P. Brown, Lieut. Comdr., USN
Frank J. Wille, Lieut. Comdr., USN
George H. Emmerson, Lt. Comdr., USN
Paul A. Stevens, Lieutenant, USN
Lawrence J. K. Blades, Lieut., USN
Lyle G. Fear, Lieut, (jg), USNRF
Edward Wenk, Ensign (T) (G), USN
John E. Burger, Ensign (T) (M), USN
Elroy G. True, Ensign (T) (M), USN
Clarence Cappel, Ensign, USNRF
Albert P. Reals, Ensign, USNRF
Donald M. Hicks, Ensign, USNRF
Clarence S. Bishop, Ensign, USNRF
John M. Griffin, Ensign, USNRF
George H. Cottrell, Ensign, USNRF
James G. Williamson, Ensign, USNNV
Clarence F. Williams, Ensign, USNRF
William T. Gill, Jr., Asst. Surgeon, USN

Charles E. Swithenback, Asst. Paymaster, USN.

Charles V. Ellis, Act. Chaplain, USN

Carl R. Lemke, Boatswain, USN

Carl W. Nelson, Machinist, USN

William English, Carpenter, USN

Alexander Riggin, Pay Clerk, USN

Thomas E. Harris, Act. Pay Clerk (T). USN

Relief Officers in Order of Reporting on Board

Elmer S. Small, Lieut, (jg), USNRF

Fred M. Byers, Ensign, USN

Lindsay Bradford, Ensign (T), USN

William L. G. Gibson, Ensign (T), USN

Edwin E. Sheridan, Ensign (T), USN

Henry C. Taylor, Ensign (T), USN

Leslie R. Lingeman, Asst. Surgeon, USN

Daniel Hunt, Surgeon, USN

George S. Arvin, Lieut, (jg), USN

Henry F. Floyd, Lieutenant, USN

R. J. Kingsmill, Lieut. (M) (jg) (T), USN

Andres J. Norgaard, Lieut., USNRF

Eugene C. Sweeny, Lieut, (jg), USNRF

Donald M. Lovejoy, Ensign, USNRF

Jay S. Rogers, Ensign, USNRF

Henry J. Price, Boatswain (T), USN

Theo. A. Small, Gunner (T), USN

Edward F. Wilson, Gunner (T), USN

Louis J. Miller, Machinist (T), USN

William F. Dickerson, Carp. (T), USN

James F. Bryant, Machinist (T), USN

Oliver E. Cobb, Ensign, USNRF

Sergius M. Riis, Lieutenant, USNRF

U. S. S. Orizaba

(Transport)

R. Drace White, Commander, USN

William P. Williamson, Lieut. Comdr., USN

W. J. Willet, Lieut. Comdr., USNRF

William D. Prideaux, Lieut. Comdr., USNRF

Philip V. H. Weems. Lieutenant, USN

John C. Tyler, Lieutenant, USN

Condie K. Winn, Surgeon, USN

Elliot Ranncy, Paymaster, USN

Edwin W. Hill, Lieut, (jg), USN

Paul Pennington. Lieut, (jg), USNRF

Robert R. Farnum. Ensign, USNRF

John E. Cutchins, Carpenter, USN

Leo L. Waite, Gunner, USN

Claude Gunn, Gunner, Ord., USN

Carston F. Olsen, Lieut., USNRF

E. B. Small, Lieutenant, USNRF

O. F, Schroeder, Lieutenant, USNRF

Charles J. Conners, Lieut. (jg), USNRF

Wm. B. Duncan, Ensign, USNRF

Alex. H. Twombly, Ensign, USNRF

Wm. W. Holton, Lieut, (jg), USN

John H. Chase, Lieut, (jg), USN

Joseph F. Meade, Lieut, (jg), USNRF

Edmond T. Coon. Ensign, USN

Henry A. Guba, Ensign, USNRF

Penn Gaskill Skillern, P.A. Surgeon, USNRF

Raymond H. Krepps, Asst. Surg., USN

Max E. Zimmerman, Pharm., USN

Robert G. Rauscher, Asst. Paymaster,USNRF

Hector J. Gosselin, Pay Clerk, USN

John B. Sloggett, Ensign, USNRF

Elliot C. Terhune, Ensign, USNRF

Kilborn B. Coe, Ensign. USNRF

Hosea B. Phillips, Ensign, USNRF

Harry E. Vercy, Lieutenant, USNRF

Relief Officers in Order of Reporting on Board

Louis A. Babcock, Lieutenant, USNRF

George D. Perry, Lieutenant, USNRF

James L. McCormack, Lieut, (jg), USN

Percy A. Cook, Lieut, (jg), USN

John E. Leary, Ensign, USNRF

Sumner A. Mead, Ensign, USNRF

Paul A. Mather, Ensign, USNRF

Hildreth Meigs, Ensign, USNRF

Dexter H. Marsh, Ensign, USNRF

Louis Lombardi, Ensign, USNRF

Carl A. R. Lewis, Ensign. USNRF

Arthur L. McCobb, Ensign, USNRF

Arthur Grove, Boatswain (T), USN

S. H. Warner, Ensign, USN

R. G. Waldron, Ensign, USN

Wallace Bertholf, Commander, USN

Henry V. Cranston, Ensign, USNRF

Edwin V. S. Boyle, Ensign, USNRF

Thomas C. Barnes, Ensign, USNRF

Norman J. Patterson, Ensign, USNWT

William V. Tubby, Ensign, USNRF

James D. Lucey, Ensign, USNRF

William H. Clinton, Gunner (T) (O), USN

Fred C. McCormack, Pay Clk., USNRF

Charles S. Freeman, Captain, USN

G. O. Olsen, Lieut, (jg) (D), USNRF

George W. Riley, Ensign (T), USN

H. H. Roberts, Ensign (T), USN
J. T. Roach, Ensign (T), USN
D. E. Robertson, Ensign (T), USN
L. E. Robinson, Ensign (T), USN
Howard Priest, Lt. Comdr. (MC), USN
Grosbeck F. Walsh, Lt. (MC), USNRF
E. J. Shaughnessy, Pay Clerk, USNRF
A. A. Bigelow, Lieutenant, USN
W. H. Beauvais, Machinist, USNRF
M. P. Ferguson, Machinist, USNRF
W. A. Stickney, Machinist, USNRF

U. S. S. Pastores

(Transport)

C. W. Cole, Commander, USN
R. G. Haxton, Lieutenant, USN
R. A. Awtrey, Lieutenant, USN
M. Comstock, Lieutenant, USN
D. Cook, Lieutenant, USNRF
J. W. Wilson, Lieutenant, USNRF
E. Rowell, Lieutenant (jg), USNRF
F. A. Mosher, Lieutenant (jg), USNRF
E. P. Shevlin, Lieutenant (jg), USNRF
M. E. Levy, Lieutenant (jg), USNRF
A. W. Bang, Lieutenant (jg), USNRF
E. Stetter, Lieutenant (jg), USNRF
M. Burke, Ensign, USN
C. H. Fogg, Ensign, USN
G. N. Maynard, Ensign, USNRF
J. M. Burke, Ensign, USNRF
W. S. Tulloch, Ensign, USNRF
S. Kohn, Ensign, USNRF
M. E. Goldstone, Ensign, USNRF
N. H. Findly, Ensign, USNRF
J. G. Enright, P. A. Surgeon, NNV
C. K. Reinke, Asst. Surgeon, USN
M. T. Briggs, Asst. Surgeon, USNRF
N. S. Trottman, Asst. Paymaster, USN
W. M. Rees, Asst. Paymaster, USNRF
J. Sargeant, Boatswain, USN
J. A. McDonough, Carpenter, USN
T. Endres, Machinist, USNRF
G. Grosch, Machinist, USN
J. Carney, Machinist, USNRF
A. F. Reed, Machinist, USNRF
F. M. Garaghty, Pay Clerk, USN
S. B. Dodson, Pharmacist, FNR
L. E. Hough, Pharmacist. USN

Relief Officers in Order of Reporting on Board

G. C. De Lacy, Ensign, USN
Edwin B. Dickinson, Ensign (T), USN
George J. Carr, Ensign (T), USN
Grover C. Elder, P. A. Surg., USNRF
Ralph Schmucker, Lieut, (jg), (DC), USNRF
Merlyn G. Cook, Commander, USN
William S. Howell, Ensign (T), USN
James L. Hinds, Ensign (T), USN
R. D. McManigal, Jr., Ensign, USNRF
Wm. S. Morgan, Lieut. (MC), USNRF
Matt Elson, Lieut, (jg), USNRF
A. W. MacNichol, Ensign, USNRF
N. J. Halpine, Ensign (PC), USNRF
Wm. F. Ahrens, Bosn. (T), USN
Wm. H. Hughes, Gunner (T), USN

U. S. S. Plattsburg

(Transport)

C. C. Bloch, Commander, USN
W. J. Roberts, Lieut. Comdr., USNRF
C. H. Boucher, Lieutenant, USN
J. O. Downey, Surgeon, USN
R. E, Dennett, Lieutenant, USN
B. S. Gants, Paymaster (Asst.), USN
J. C. Taylor, Asst. Surgeon, USN
G. C. Cartmell, Lieutenant, USNRF
W. J. Munroe, Lieutenant, USNRF
C. Fournier, Lieutenant, USNRF
G. S. Mundie, Lieutenant, USNRF
A. E. Harding, Lieutenant, USNRF
E. Mullaley, Asst. Surgeon, USN
C. E. Scouller, Lieut, (jg). USNRF
T. Burgess, Lieut. (jg) USNRF
J. C. Meyers, Lieut. (jg) USNRF
H. Hammond, Lieutenant (jg), USNRF
J. R. Spear, Lieut, (jg), USNRF
J. M. Handley, Lieut, (jg), USNRF
E. Lewis, Lieut, (jg), USNRF
C. Isgarr, Lieut, (jg), USNRF
T. Stevenson, Asst. Paymaster, USNRF
W. W. Hedges, Ensign, USN
P. Dean, Asst. Paymaster, USNRF
E. R. Walsh, Ensign, USNRF
F. V. Greene, Ensign, USNRF
S. Wincapaw, Ensign, USNRF
P. E. Perez, Ensign, USNRF
W. Jones, Ensign, USNRF
C. W. Jackson, Ensign, USNRF

W. Herlihy, Ensign, USNRF
L. E. Orvis, Ensign, USNRF
C. J. Muhlfeld, Ensign, USNRF
P. E. Aldrich, Ensign, USNRF
K. S. Smith, Ensign, USNRF
R. M. Gotham, Ensign, USNRF
W. T. Smart, Boatswain, USN
B. L. Wood, Carpenter, USN
J. E. Baum, Pharmacist (T), USN
E. J. Hoffman, Pay Clerk, USNRF
W. H. Quayle, Carpenter, USNRF
A. N. Gale, Pay Clerk, USNRF
W. Barlow, Carpenter, USNRF
W. Neumeyer, Gunner (T), USN
W. Freemantle, Machinist, USNRF
W. Gallup, Machinist, USNRF
F. W. Southworth, Asst. Pann., USNRF
John A. Nicol, Carpenter (T), USN
P. F. Kennedy, Dental Surgeon, USN
Donald S. Tuttle, Ensign, USNRF
William E. Murphy, Ensign, USNRF
Joseph A. Mc Vicar, Ensign, USNRF
Fred B. Smith, Asst. Surgeon (T), USN
Benj. C. Britt, Carpenter (T), USN

*Relief Officers in Order of Reporting on
Board*

T. M. Burke, Lieut, (jg), USNRF
W. H. O'Donaghue, Lieut, (jg), USNRF
R. L. Armstrong, Ensign, USNRF
J. F. Bryan, Ensign, USNRF
W. B. liarckerman. Ensign, USNRF
E. G. Martin, Ensign, USNRF
L. R. Madison, Ensign, USNRF
H. H. Luedinhaus, Ensign. USNRF
W. R. Hughes, Ensign, USNRF
P. L. Hughes, Ensign, USN
W. J. Hudson, Ensign, USNRF
H. L. Hudson, Ensign, USNRF
N. Hoag, Ensign, USNRF

U. S. S. Pocahontas
(Formerly PRINCESS IRENE) (Transport)

J. F, Hellweg, Commander, USN
Burton H. Green, Lieut. Comdr., USN
James L. Oswald, Lieutenant, USN
James W. Fleming, Lieutenant, USNRF
Walter P. Raarup, Lieut, (jg), USNRF
Willis C. Sutherland, Ensign, USN
Samuel L. Wartman, Ensign, USN
John H. O'Leary, Ensign, USNRF

Charles F. Adae, Ensign, USNRF
Walter D. Guiney, Ensign, USNRF
David J. Laraie, Ensign, USNRF
Micajah Boland, Asst. Surgeon, USN
Paul A. Clark, P. A. Paymaster, USN
Reuben L. Larsen, Asst. Surg., USNRF

*Relief Officers in Order of Reporting on
Board*

A. B. Johnson, Elec. Gunner, USNRF
Frank Steele, Machinist, USNRF
Frederick G. Legere, Boatswain, USN
W. J. Graham, Gunner, USN
G. D. Barringer, Carpenter, USN
G. G. Holton, Asst. Surgeon, USNRF
W. C. Gray, Chief Machinist, USN
E. L. Newell, Gunner, USN
F. E. Herbert, Pay Clerk, USN
H. B. Fluck, Lieutenant (jg), USN
A. W. Hinckley, Ensign, USN
H. D. Nuber, Asst. Paymaster, USN
E. B. Ericcson, Asst. Paymaster, USN
W. R. Thomas, Asst. Naval Constr., USN (T)
W. H. Stuart, Machinist (T), USN
F. C. Lemke, Machinist, NNV
W. J. Bisel, Pharmacist, USN
D. M. Wood, Commander, USN
W. F. Sellers, Asst. Paym.. USNRF
E. C. Kalbfus, Commander, USN
W. E. Morrison, Lieut, (jg), USNRF
J. C. Acvedo, Ensign, USNRF
A. W. Dixon, Ensign, USNRF
R. L. Koester, Pay Clerk, USN
J. J. McMahon, Carpenter (T), USN
R. Agerup, Lieutenant, USNRF
P. C. Cornelius, Lieut, (jg) USNRF
A. F. Foss, Lieutenant (jg), USNRF
V. G. Clark, P. A. Surgeon, USNRF
E. L. Ackiss, Acting Chaplain, USNRF
B. H. Barton, Dentist, USNRF
J. D. Eggleston, Ensign, USNRF
H. T. Keyes, Ensign, USNRF
L. V. Klauberg, Ensign, USNRF
J. G. S. Humphreys, Ensign, USNRF
R. P. Hughes, Ensign, USNRF
R. C. Farnham, Ensign, USNRF
W. J. Strachan, Ensign, USN
W. J. Carter, Asst. Paymaster, USN
B. F. Iden, P. A. Surgeon, USNRF
W. Glenn, Asst. Paymaster, USNRF
L. E. Bratton, Lieut. Comdr., USN
W. A. Spencer, Boatswain, USN

W. L. Saunders, Asst. Paym., USNRF
C. A. Krez, Lieutenant, USN
W. W. Jones, Lieutenant (MC) USNRF
B. F. Andrews, Lieut. (jg) (MC), USNRF
B. G. Holton, Lieut. (MC), USN
F. P. Moore, Machinist, USN
E. Guthrie, Lieut. Commander, USN
R. B. Parker, Ensign (PC), USNRF
J. O. Jenkins, Ensign (T). USN
T. H. Hunter, Ensign (T), USN
W. H. Jones, Ensign (T), USN
C. F. Jacobsmeyer Ensign (T), USN
M. P. Hall, Ensign (T), USN
E. G. Metcalf, Ensign (T), USN
R. I. Mayorga, Ensign (T), USN

U. S. S. Powhatan

(Transport)

G. S. Lincoln, Commander, USN
W. W. Smyth, Lieut. Commander, USN
J. W. Hayward, Lieut. Comdr., USN
R. T. Young, Lieutenant, USN
F. C. McMurry, Lieutenant, USNRF
G. G. Berwind, Lieutenant (jg), USN
F. A. Finch, Lieutenant (jg), USN
N. A. Bolin, Lieutenant (jg), USN
E. A. Peter, Lieutenant (jg), USNRF
F. W. Davis, Lieutenant (jg), USNRF
T. E. Chapman, Lieutenant (jg), USNRF
L. G. Smith, Lieutenant (jg), USNRF
B. F. Singles, Ensign (T), USN
U. G. Chipman, Ensign (T), USN
H. G. Oliver, Ensign (T), USN
F. J. Murphy, Ensign (T), USN
F. W. Yurasko, Ensign (T), USN
E. E. Curtis, P. A. Surgeon, USN
R. H. Johnson, P. A. Surgeon, USN
W. C. Becker, Asst. Surgeon. USN
C. W. Eley, Asst. Surgeon, USN
R. M. Anderson, Asst. Paym., USN
J. W. Sprague, Asst. Paymaster, USN
H. W. Crider, Asst. Paym. (T), USN
P. K. Coons, Pay Clerk (Act.), USN
W. T. Minnick, Pharmacist, USN
W. Collins, Carpenter, USN
W. Lovell, Machinist, USN
S. C. Harrison, Ensign, USNRF
T. E. Mason, Ensign, USNRF
H. W. Waugh, Ensign, USNRF
J. A. Ryan, Ensign, USNRF
J. O. Callender, Ensign, USNRF

T. E. Flaherty. Ensign (T), USN
J. F. O'Brien, Carpenter, USN
G. E. Lenski, Chaplain, USN
C. Ivins, Machmist, USN
R. Chaney, Gunner, USN
N. C. Lovegrove, Lieut. (jg), USNRF
E. H. Proudfit, Ensign, USNRF
W. L. Curry, Ensign, USNRF
W. B. Bryant. Ensign, USNRF
T. O. Helm, Ensign, USNRF
R. B. Russell, Ensign, USNRF
F. L. Healy, Ensign. USNRF
E. A. Daus, Asst. Surgeon, USN
R. B. Blackwell, Asst. Surg., USNRF
C. B. Somers, Asst. Paymaster, USNRF
N. S. Mack. Ensign (t), USN
N. G. McKee, Ensign (T), USN

Relief Officers in Order of Reporting on Board

J. P. Murdock, Commander, USN
R. A. Gilbert, Asst. Surgeon, USN
R. W. Miller, Asst. Paymaster, USN
W. P. McNamara, Ensign, USNRF
J. A. Flint, Ensign, USNRF
D. W. Loomis, Lieutenant (jg), USN
F. Falkenstein, Lieutenant (jg), USN
W. Teeuwe, Machinist (T), USN
P. K. Coons, Pay Clerk (Act.), USN
C. Freund, Lieutenant (jg), USNRF
J. B. Clapp, Ensign, USN
P. S. Mock, Ensign (T), USN
D. F. Wilson, Ensign (T), USN
O. Rhode, Boatswain, USN
O. M. Southard, Ensign, USNRF
C. F. Bauman, Ensign, USNRF

U. S. S. President Grant

(Transport)

J. P. Morton, Commander, USN
W. W. Galbraith, Lieut. Comdr., USN
S. S. Brown, Lieutenant, USN
H. W. Hosford, Lieutenant, USN
Wm. S. Pearson, Lieutenant, USNRF
R. F. Walter, Lieutenant, USNRF
F. E. Coops, Lieutenant (jg). USNRF
J. Kronholm, Lieutenant (jg), USNRF
M. A. Kerr, Lieutenant (jg). USNRF
G. C. Bartlum, Lieutenant (jg), USNRF
J. S. Waters, Ensign, USN
N. R. George, Ensign, USN

P. O. McDonough, Ensign, USNRF
H. A. May, Surgeon, USN
M. B. Miller, Surgeon, USNRF
W. L. Martin, Asst. Surgeon, USN
J. R. Allison, Asst. Surgeon, USNRF
F. W. Holt, Paymaster, USN
Jos. Sperl, Boatswain, USN
A. V. Watson, Gunner, USN
A. W. Hinman, Gunner, USN
T. T. Emerson, Machinist, USN
L. E, Prey, Carpenter, USN
R. C. Rowe, Pharmacist, USN
T. M. Smith, Pay Clerk, USN

*Relief Officers in Order of Reporting on
Board*

H. F. Jennings, Ensign, USNRF
C. Cappell, Ensign, USNRF
J. N. Campbell, Asst. Paym., USNRF
Paul Buhlig, Asst. Paymaster, USNRF
R. M. Nctz, Asst. Paymaster, USN
E. E. Bell, Asst. Paymaster, USN
T. F. Deylin, Act. Pay Clerk, USN
C. T. Flannery, Act. Pay Clerk, USN
J. H. Smith, Lieutenant, USN
R. H. Allen, Lieut. Comdr., USNRF
A. L. Morgan, Jr., Lieutenant, USN
T. Murray, Lieutenant, USNRF
I. L. Church, Ensign, USNRF
W. C. Huck, Ensign, USNRF
C. E. Courtney, Commander, USN
G. C. Moses, Commander, USNRF
J. Stolan, Lieutenant, USNRF
H. M. Roberge, Lieutenant, USNRF
R. F. Gilley, Lieutenant (jg), USNRF
J. D. O'Connor. Ensign, USNRF
G. Pellegriue, Ensign, USNRF
H. D. Van Houten, Ensign, USNRF
C. R. HafFendcn, Ensign, USNRF
E. B. McElrov, Ensign, USNRF
M. Cooper, Jr.. Ensign, USNRF
D. Goldberg, Acting Chaplain, USN
B. O. Kilroy, Acting Pay Clerk, USN
C. F. McKelvey, Lieut, (jg), USNRF
J. P. Paul, Carpenter, USN
S. P. Vaughn, Asst. Paymaster, USN
J. K. Hollowell, Pharmacist, NNV
H. W. D. Rudd, Ensign (T), USNRF
D. A. Green, Lieutenant, USNRF
A. H. Dodge, P. A. Surgeon, USN
G. H. E. Robinson, 2nd Lieut., USQMC
G. C. Bartlum, Lieutenant, USNRF

P. B. Thompson, Ensign, USN
W. E. Tarbutton, Ensign, USN
J. A. Nelson, Ensign, USNRF
J. Montgomery, Ensign, USNRF
J. J. Murphy, Ensign, USNRF
Leonard Opdycke, Ensign, USNRF
E. Johnson, Ensign, USNRF
J. F. Peck, Ensign, USNRF
C. D. Langhorne, Surgeon, USNRF
F. D. Newbarr. Asst, Surgeon, USNRF
A. A. Newbarr, Asst. Surgeon, USNRF
H. B. Jablow, Asst. Surgeon, USNRF
L. A. Willard, Dental Surgeon, USN
S. L. Maxwell, Acting Chaplain, USN
A. M. Hinman, Lieut, (jg) (T), USN
T. T. Emerton, Lieut, (jg) (T), USN
J. A. Kerney, Pay Clerk, USNRF
J. F. Cremens, Lieut, (jg) (MC), USN
W. W. Lightner, Boatswain (T), USN
K. R. Pitcher, Gunner (T), USN
Johan Svesson, Gunner (T), USN
E. L. Norton, Pay Clerk, USNRF
C. W. Cole, Captain, USN
G. D. Callaway. Lieut. (MC), USN
R. D. Sample. Lieut. (MC), USNRF
W. B. Kerr, Ensign (T), USN
J. D. Kennedy, Ensign (T), USN
G. F. Metz, Ensign, USNRF
M. J. Jukich, Ensign (T), USN
D. II. Kane, Ensign (T), USN
L. P. Kane, Ensign (T), USN
E. R. Baker, Pharmacist (T), USN

U. S. S. President Lincoln
(Transport)

Yates Stirling, Jr., Captain, USN
P. W. Foote, Commander, USN
W. D. Owens, Surgeon, USN
J. W. Browning, Paymaster, USN
L. W. Lind, Lieutenant, USN
J. B. Odendorf, Lieutenant, USN
D. H. Blellock, Lieutenant, USNRF
E. v. M. Isaacs, Lieutenant (jg), USN
J. F. Donahue, Asst. Surgeon, USN
J. W. Troxell, Asst. Surgeon, USN
E. E. Merrell, Lieutenant (ig), USNRF
J. W. Willett, Lieutenant (jg), USNRF
G. F. Wells, Lieutenant (jg), USNRF
I. H. Mettern, Lieutenant (jg), USNRF
D. F. Luby, Lieutenant (jg), USNRF
Edward Baker, Lieutenant (jg), USNRF

J. W. Kirschner, Lieut, (jg), USNRF
Wm. S. Rhoades, Asst. Paym., USN
J. F. Loba, Asst. Paymaster, USN
J. E. Cleary, Ensign, USN
William Seach, Ensign, USN
C. E. Briggs, Ensign, USN
J. Shottroff. Ensign, USN
F. A. Brewer, Ensign, USNRF
E. W. Mott, Ensign, NNV
T. V. Corey, Ensign, NNV
S. J. Curry. Ensign, USNRF
G. H. Cottrell, Ensign, NNV
N. A. Winquist, Ensign, USNRF
Fred A. Just, Ensign, USNRF
A. W. Matthews, Ensign, USNRF
G. B. Kimberly, Pay Clerk, USN
J. B. Burke, Pay Clerk (T), USN
E. B. Berkstresser, Carpenter, USN
C. E. Snider, Pharmacist, USN
P. Troy, Carpenter, USNRF
R. C. Jones, Boatswain, USNRF
L. H. French, Pharmacist, USN

*Relief Officers in Order of Reporting on
Board*

William Fleming, Lieutenant, USNRF
Alan C. Blanding, Ensign, USNRF
Wesley S. Block, Jr., Ensign, USNRF
Wesley C. Martin, Ensign, USNRF
Alexander Murray, Jr., Ensign, USNRF
W. C. Manley, Chief Gunner, USNRF
A. G. Velton, Lieutenant, USNRF
Wm. T. Davidson, Dent. Surg., USNRF
J. D. Blackwood, Jr., Asst. Surgeon, USNRF
Guy Fish, Asst. Surgeon, USNRF
L. C. Whiteside. Surgeon, USN
Andrew Mowat, Asst. Paym., USN
G. C. Whimsett, Chapl. (Lt.) (jg), USN
F. B. Mullen, Lieutenant (jg), USNRF
J. R. Fairbanks, Lieutenant (E), USNRF
J. E. Johnston, Asst. Paym., USNRF
J. E. Gainard, Ensign, USNRF
R. S. Hammond, Ensign, USNRF
John S. Hill, Ensign, USNRF
A. L. Arnold, Jr., Asst. Surg., USNRF
Clinton R. Black, Jr., Ensign, USNRF

U. S. S. Princess Matoika
(Transport)

William D. Leahy, Commander, USN
H. C. Gearing, Jr., Lieut. Comdr., USN

C. G. Halpine, Lieutenant, USN
W. S. Case, Lieutenant, USN
L. C. Weith, Lieutenant, USNRF
R. H. Quynn, Lieutenant, USNRF
T. C. Bruce, Lieutenant (jg), USNRF
V. E. Anderson, Lieut, (jg), USNRF
C. A, Wagner, Lieutenant (jg), USNRF
I. D. Eby, Lieutenant (jg), USNRF
T. P. Kane, Ensign, USN
R. R. Clegg, Ensign, USNRF
J. B. Hunziker, Ensign, USNRF
G. W. Wylie, Ensign, USNRF
T. A. Waage, Ensign, USNRF
L. F. Oliver, Ensign, USNRF
B. M. Hackley, Ensign, USNRF
J. B. Naugle, Ensign, USNRF
B. M. Kendall, Ensign, USNRF
J. Hamilton, Ensign, USNRF
F. W. S. Dean, Medical Insp., USN
H. L. Kennedy, Asst. Surgeon, USN
G. B. Storey, Asst. Surgeon, USNRF
D. B. Kirby, Asst. Surgeon, USN
P. C. Corning, P. A. Paymaster, USN
C. B. Kitchen, Asst, Paymaster, USNRF
A. Page, Asst. Paymaster, USNRF
G. A. Stevens, Pay Clerk, USN
J. W. Eyers, Pay Clerk, USNRF
J. M, Kiernan, Carpenter, USN
G. W. Williams. Gunner (O), USN
E. J. McCarthy, Gunner (E), USN
C. R. Holmes, Pharmacist (T), USN
P. J. Fitzgibbons, Machinist, USN

*Relief Officers in Order of Reporting on
Board*

C. T. Owens, Commander, USN
F. B. Orr, Lieutenant, NNV
H. M. Tickle, Lieutenant (jg), USNRF
H. Redfield, Asst. Paymaster, USNRF
H. N. Lambert, Lieutenant (jg), USN
J. W. Farrell, Dental Surgeon, USNRF
P. W. O'Brien, Machinist (T), USN
W. E. Redfern, Carpenter (T), USN
R. J. Joers, Lieutenant (D), USNRF
C. Hollinshed, Lieut. (D), (jg) USNRF
P. Hemdon, Lieut. (D), (jg) USNRF
L. A. Gorman, Ensign (D), USNRF
E. W. Woolard, Ensign (D), USNRF
H. L. Fish, Asst. Surgeon, USNRF
T. A. Kittinger, Commander, USN
J. T. Steward, Ensign (E), USNRF
L. T. Lavalley. Ensign (T), USN

W. A. Kingsbury, Ensign (T), USN
F. E. Kyle, Ensign (T). USN
W. M. Klein, Ensign (T), USN
T. E. Lake, Ensign (T), USN
H. W. Moss, Ensign (T), USN
J. F. Murphy, Ensign (T), USN
S. F. Strong, Pharmacist (T), USN

U. S. S. Pueblo
(Cruiser)

George W. Williams, Captain, USN
Manley H. Simons, Commander, USN
William A. Glassford, Jr., Lieut. Comdr. USN
William C. Owen, Lieutenant, USN
John F. McClain, Lieutenant, USN
August Schulze, Lieutenant, USN
Oliver H. Ritchie, Lieutenant, USN
Carlyle Craig, Lieutenant (jg), USN
Herbert W. Anderson, Lieut, (jg), USN
William Busk, Lieutenant (jg). USN
Willis M. Pcrcifield, Lieut, (jg), USN
Thomas P. Clark, Lieut, (jg), USN
Andrew N. Anderson, Ensign (T) (B), USN
Frank E. Nelson, Ensign (T) (M), USN
Thomas J. Sullivan, Ensign (T) (M), USN
Wilmer W. Weber, Ensign (T) (M), USN
Omar B. Earle, Ensign (T) (G), USN
Jay Smith, Ensign (T) (B), USN
Julian T. Miller, P. A. Surgeon, USN
Foster H. Bowman, Asst. Surgeon, USN
Emory D. Stanley, Paymaster, USN
Harold C. Pierce, Captain, USMC
Evert O. Smith, Carpenter, USN
Charles A. Sieck, Ac-t. Pay Clerk, USN
Harry Bennett, Machinist (T), USN
Marvin McCray, Act. Pay Clerk (T), USN
Arthur R. Wallen, Act. Pay Clerk (T), USN
Jesse II. Porth, Act. Pay Clerk (T), USN
Wayne D. Thompson, Lieutenant (jg), US-NRF
Ravmond C. Darrow, Lieutenant (jg) USNRF
Douglas G. Lovell. Ensign (T), USN
James H. Willey, Lieutenant, NNV
Frank R. Seaver, Lieutenant, NNV
Frank S. M. Harris, Lieutenant, NNV
William C. Tooze, Lieutenant, NNV
David A. Loebenstein, Lieut., NNV
Henry C. Buckle, Lieut, (jg), NNV
Ralph J. A. Stern. Ensign, NNV
Rudolph T. Haas, Ensign, NNV

Alfred A. Marietta, P.A. Surgeon, NNV

Relief Officers in Order of Reporting on Board

A. O. Kolstad, Ensign (T), USN
H. F. McGee, Ensign (T), USN
E. C. Fugler. Ensign, USNRF
L. H. Denny, Asst. Surgeon, USN
P. A. Caro, Asst. Paymaster, USN
C. A. Milliken, 2nd Lieutenant, USMC
E. R. James, Ensign, USNRF
P. H. Powers, Ensign, USNRF
D. B. Flood, Ensign. USNRF
R. L. Whitcomb, Ensign, USNRF
G. da S. Nunes, 1st Lieut. Braz. Navy
A. da F. Costa, 1st Lieut. Brazil. Navy
A. Vidal. Ensign. USNRF
J. M. Hester. Chaplain, USNRF
Eric M. Grimsley, Ensign, USN
John B. Griggs, Ensign, USN
John A. Cloyd, Asst. Surgeon, USN
Lewis A. Francis, Asst. Surg. (D), USN
Raymond V. Christmas, Pay Clk., NNV
B. H. Mack. Lieutenant (jg), USN
Benjamin Dutton, Jr., Comdr., USN
William J. Nunnallv, Jr., Lieut., USN
Lloyd E. Clifford, Lieut, (jg). USN
William Shear, Carpenter, USN
Willie M. Kenyon, Gunner, USN
Hiram O. Hartley, Gunner, USN
Frank B. Upham, Captain, USN
Frederick L. Ryon, Ensign (T), USN
Wm. H. Hopkins, Act. Pay Clerk (T), USN
Frederick J. Collins. Machinist (T), USN
John E. Quint, Machinist (T), USN
Wilfred G. Conrad, Machinist, USNRF
Edgar R. McClung, Lieut. Comdr., USN
Richard C. English, Ensign, USNRF
Walter B. Holder, Ensign (T), USN
John A. Mayer, Ensign (T), USN
John R. Matthews, Ensign (T), USN
E. A. Magill, Ensign (T), USN
Ralph S. Maugham, Ensign (T), USN
Norman E. Miller, Ensign (T), USN
Ransdell Matthews, Ensign (T), USN
Francis G. Minor, Ensign (T), USN
S. N. Minor, Ensign (T), USN
William Van D. Jewett, Captain, USMC

U. S. S. Raleigh

(Cruiser)

C. J. Lang, Commander, USN
F. D. Pryor, Lieutenant, USN
F. J. Lowry, Lieutenant (jg), USN
E. M. Zacharias, Lieutenant (jg), USN
F. M. Harrison, Asst. Surgeon, USN
H. E. Burke. Ensign. USN
G. P. Martin, Ensign. USN
F. S. Gibson, Ensign, USN
H. D. Stailey, Ensign, USN
E. D. Gibb, Ensign, USN
T. C. Gibbs, Asst. Paymaster, USN
R. B. Deming, Chief Pay Clerk, USN
W. EUiott, Pay Clerk (t), USN

Relief Officers in Order of Reporting on Board

W. E. Glanville, Asst. Surgeon, USN
R. S. Maynard, Ensign, USNRF
W. H. May, Ensign, USNRF
D. G. Lovell, Ensign, USNRF
D. McClench, Ensign, USNRF
J. Ball, Pay Clerk, (T) USN
C. J. Wacker, Pay Clerk (T), USN
E. G. Ross, Pay Clerk (T), USN
C. C. Stailey, Gunner (T), USN
J. M. Groff, Carpenter (T), USN
F. E. Ridgeley. Commander, USN
P. C. Moyer, Ensign, USNRF
G. M. Pulver, Ensign, USNRF
W. G. Pritchard, Ensign, USNRF
E. Nurenberg, Ensign, USNRF
L. W. Morgan, Ensign, USNRF
J. J. Shipherd, Ensign, USNRF
A. O. Siler, Ensign, USNRF
B. A. Smith, Ensign, USNRF
W. D. Thomas, Lieutenant, USN
R. B. Kellogg, Lieut, (jg), USNRF

U. S. S. Rijndam

(Transport)

J. J. Hannigan, Commander, USN
S. S. Payne, Lieut. Commander, USN
W. K. Martin, Lieut. Comdr., USNRF
B. K. Presnell, Lieutenant, USN
S. Greenlee, Lieutenant, USNRF
C. H. T. B. Tisell, Lieutenant, USNRF
H. Arneson, Lieutenant (jg), USNRF

W. J. Carr, Lieutenant (jg), USNRF
W. J. Hantel, Lieutenant (jg), USNRF
P. V. Lane, Lieutenant (jg), USNRF
J. H. Nolan, Lieutenant (jg), USNRF
O. Olsen, Lieutenant (jg), USNRF
R. E. Berlin, Ensign, USNRF
R. H. S. Booth, Ensign, USNRF
C. E. Carter, Ensign, USNRF
A. J. M. Grant, Ensign, USNRF
F. J. Kasper, Ensign, USNRF
L. R. Loney, Ensign, USNRF
A. R. Murray. Ensign, USNRF
F. R. Neindorff, Ensign, USNRF
W. J. Slattery, Ensign, USNRF
E. H. Tinker, Ensign, USNRF
William Hardy, Boatswain, USN
A. A. Franks, Gunner (T), USN
W. A. Gerdts, Gunner (T). USN
R. C. Pomerov, Machinist, USN
A. D. McGillvrav, Carpenter (T), USN
P. Troy, Carpenter, USNRF
E. A. Vickery, Surgeon, USN
W. B. Anderson, Asst. Surgeon, USN
W. M. Brunet, Asst. Surgeon, USNRF
S. S. Gant, Pharmacist, LSN
W. G. Swearingen, Asst. Paym., USN
P. D. Benner, Asst. Paymaster, USNRF
W. G. Springer, Pay Clerk, USNRF
Louis Kloker, Lieutenant (jg), USNRF
V. Carroll, Pay Clerk (T), USN
A. R. Walsh, Asst. Paymaster, USNRF
R. D. Earle, Asst. Paymaster, USNRF

Relief Officers in Order of Reporting on Board

T. Voss, Lieutenant, USNRF
Wm. B. Lockwood, Ensign, USNRF
A. W. Bachman, Ensign, USNRF
H. C. Shepheard, Ensign, USNRF
H. J. Benzoni, Ensign, USNRF
J. P. Wright, Jr., Ensign, USNRF
J. R. Warren, Ensign, USNRF
Edward J. Fitzgerald, Asst. Surgeon, USNRF
Harry F. Horan, Lieutenant, USNRF
G. G. Robertson, Lieutenant, USN
H. F. Melching, Ensign (E), USNRF
C. M. Stevenson, Lieut, (jg). USNRF
W. L. Loekwood, Ensign (D), USNRF
L. F. Dowrie, Ensign (D), USNRF
F. A. Comstock, Ensign (D), USNRF
F. R. Koppen, Ensign (D), USNRF
R. S. Haley, Ensign, USNRF

D. Y. Wemple. Ensign, USNRF
M. F. Walsh, Ensign, USNRF
T. J. Van Tmsk, Ensign, USNRF
R. C. Ziegler, Ensign, USNRF
G. E. Seaman, Ensign (E), USNRF
L. T. Hopkins, Asst. Surgeon, USN
F. M. Conrad, Asst. Paymaster, USN
Edward Fief, Ensign, USNRF
S. B. Sawtelle, Ensign, USNRF
J. S. Salom, Ensign, USNRF
C. R. Sanders, Ensign, USNRF
E. W. Roemer, Ensign, USNRF
L. S. Walsh, Ensign, USNRF
Allen S. Noyes, Ensign, USNRF
James Rolchford, Gunner (T). USN
J. R. Coffey. Lieut, (jg), USNRF

U. S. S. Rochester

(Cruiser)

A. W. Hinds, Captain. USN
H. O. Roesch, Lieutenant, USN
C. S. Graves, Lieutenant, USN
J. James, Lieutenant, USN
R. H. Wakeman, Lieutenant, USN
J. G. Moyer, Lieutenant, USN
W. E. McNelly, Lieutenant (jg). NNV
A. H. Hawley. Ensign (T), USN
C. R. Doll, Ensign (T). JSN
O. E. Reh, Ensign (T), USN
S. V. Dunham. Pay Clerk, USN
A. F. Benzon, Ensign (T), USN
R. M. Munson, Carpenter, USN
A. J. Geiger, Surgeon, USN
L. J. Wolf, P. A. Surgeon, USN
A. L. Hodgson, Ensign. NNV
G. R. Snider, Ensign, NNV
J. H. Colhoun, P. A. Paymaster, USN
R. S. Hatch, Lieutenant (jg), USN
H. S. Jones, Lieutenant (jg). USN
E. J. Hodgdon, Lieutenant (jg), USN
E. H. Price, Ensign, USN
R. F. Gray, Machinist (T), USN
G. B. Crow, P. A. Surgeon, USNRF
L. H. Larson, Carpenter, USNRF
C. H. Gordon, Gunner (E) (T), USN
W. P. Boardman, Machinist (T), USN
A. J. McDaniel, Act. Pay Clk. (T), USN
C. W. Jordan, Machinist (T), USN
G. F. Murphy, Asst. Chaplain, USN
F. C. Vossbeck, Dental Surgeon, USN
S. L. Scott, Asst. Surgeon, USN

Relief Officers in Order of Reporting on Board

S. W. Burton, Ensign (T), USN
K. A. Diechman, Lieut, (jg), USNRF
C. C. MacDougall, Act. Pay Clerk (T), USN
A. D. Holland, Boatswain (T), USN
T. D. Shepherd, Ensign, USNRF
E. M. Jaeger, Ensign, USNRF
N. L. Fortin, Ensign, USNRF
F. K. Williams, Lieut, (jg), USNRF
W. E. Holland, Lieut, (jg), USNRF
R. S. Elder, Ensign, USNRF
J. J. Fitzgerald, Ensign, USNRF
J. R. Tobin, Ensign, USNRF
G. W. Ayer, Ensign (T), USNRF
W. E. Scott, Asst. Paymaster, USNRF
J. N. Saul. Machinist, USN
F. C. Sammons, Ensign, USNRF
H. Shortall, Ensign, USNRF
S. Parrish, Gunner, USN
E. C. Sorenson, Pay Clerk USN
D. C. King, Ensign, USN
J. E. Kiernan, Ensign, USN
H. V. Cornett, Asst. Surgeon, USN
F. K, Williams, Lieutenant (jg), USN
A. W. Robins, Pay Clerk, USNRF
H. H. Eliassen, Boatswain (T), USN
J. N. Smith, Boatswain (T), USN
Frederick Kidstrom, Bosn., USNRF
James C. Acford, Mach. (T), USNRF
R. D. MacMurdy, Ensign. USNRF
E. A. Jordan, Ensign, USNRF
R. G. Megargel, Ensign, USNRF
G. R. Paradise, Ensign (T), USN
H. N. Paradise, Ensign (T), USN
R. C. Mould, Ensign (T), USN
J. R. Morton, Ensign (T), USN
S. H. Oviatt, Ensign (T), USN
R. J. Mailhouse, Ensign (T), USN
E. H. Lewis, Ensign (T), USN
A. W. O'Connell, Ensign (T), USN
D. G. O'Connor, Ensign (T), USN
A. S. Reid, Pay Clerk, USN
E. T. Constien, Captain, USN
H. R. Partridge, Ensign, USN
C. N. Carver, Lieut, (jg), USNRF
Paul Burt (Chapl.), Lieut, (jg), USN
E. C. Pundt, Gunner (T) (O), USN
C. H. Truellinger, Carpenter (T), USN
W. P. Boardman, Ensign (M), USN

U. S. S. St. Louis

(Cruiser)

M. E. Trench, Commander, USN
G. Whitlock, Lieut. Commander, USN
R. L. Stover Lieutenant USN
F. A. L. Vossler Lieutenant, USN
S. W. King, Lieutenant (jg), USN
R. A. Hall, Lieutenant (jg), USN
W. E. Cheadle, Lieutenant (jg). USN
J. P. Brown, Ensign, USN
M. W. Powers. Ensign, USN
R. W. Clark, Ensign, USN
W. D. Austin, Ensign, USN
C. N. Fiske, Surgeon, USN
R. B. Lupton, Paymaster (Ret.), USN
C. R. Sies, Asst, Paymaster (T), USN
E. E. Carr, Asst. Surgeon, USN
J. V. McAlpine, Dental Surgeon, USN
H. W. Lyon, Lieutenant (jg), USNRF
R. W. Clark, Lieutenant, NNV
G. E. Link, Lieutenant, NNV
J. L. Armstrong, Lieutenant, NNV
N. Taylor, Lieutenant, NNV
J. A. McKeown, Lieutenant (jg), NNV
C. W. Wright, Lieutenant (jg), NNV
H. K. Koebig, Lieutenant (jg), NNV
H. L. Killer, Ensign, NNV
M. J. Hageman, Ensign, NNV
H. S. Ryerson, Ensign, NNV
H. W. Engel, Ensign, NNV
S. W. Tay, Ensign, NNV
F. C. Wisker. Ensign (T), USN
C. Dunne, Ensign (T), USN
S. E. Guild, Jr., Ensign, USNRF
G. S. Silsbee, Ensign, USNRF
W. F. Olson, Ensign, USNRF
J. L. Rothery, Ensign, USNRF
L. E. Burwell, Ensign, USNRF
H. G. Millington, Ensign, USNRF
E. T, Hammond, Boatswain, USN
E. L. Newell, Gunner, USN
G. Growney, Chief Machinist, USN
E. V. Hand, Machinist, USN
R. J. Leahy, Carpenter, USN

Relief Officers in Order of Reporting on
Board

L. D. Webb, Gunner, USN
N. King, Asst. Surgeon, USN
L. McIntyre, Boatswain (T), USN

W. T. Meyer, Gunner (T), USN
W. Evans, Captain, USN
W. W. Edel, Chaplain (Acting), USN
J. Q. Adams. Captain, USMC
D. R. Fox, 1st Lieutenant, USMC
L. F. Basse, Machinist, USN
E. P. Nolan, Ensign, USNRF
C. E. Kieser, Ensign, USNRF
O. P. Shattuck, Ensign, USN
F. H. Wight, Ensign, USNRF
C. K. Smyth, Act. Pay Clerk, USN
L. E. Clifford, Lieutenant (jg), USN
E. C. Jackson, Ensign, USN
J. P. Burlingham, Ensign, USN
V. C. Bixby, Ensign, USN
V. P. Kaercher, Ensign, USN
J. F. Robbins, Lieutenant (jg), USNRF
L. Stock, Jr., Ensign, USNRF
A. C. Stevens, Ensign, USNRF
D. M. Taylor. Ensign, USNRF
L. C. De Veaux, 2nd Lieutenant, USMC
C. A. Morton. Pay Clerk, USNRF
M. L. Weissberger. Dental Surg., USN
C. C. Rounds. Ensign (NE). USN
S. E. Guild. Jr.. Ensign, USNRF
F. A. Dixon, Ensign, USNRF
E. D. Kern, Midshipman, USN
D. Kiefer, Midshipman, USN
Amon Bronson, Captain, USN
C. J. Culbert, Ensign, USNRF
R. G. Warren, Ensign, USNRF
R. L. Williams, Ensign, USNRF
F. S. Woodruff, Ensign, USNRF
G. S. Lincoln, Captain, USN
F. G. Kutz, Lieutenant, USN
J. J. Shipley, Ensign, USNRF
J. D. Ryall. Ensign, USNRF
R. T. Bookmyer, Ensign, USNRF
S. A. Fuqua, Lieutenant (MC), USN
J. A. Poulter, Captain, USMC
G. S. Silsbee, Lieutenant (jg), USNRF
J. L. Taylor, Ensign, USN
T. J. Taylor, Ensign, USN
A. J. Storm, Ensign, USN
T. H. Snyder, Ensign, USN
J. K. Stevenson, Ensign, USN
H. E. Small, Ensign, USNRF
H. F. Parks, Ensign, USNRF
S. L. Oliver, Ensign, USN
T. Nelson, Lieut. Commander, USNRF

U. S. S. St. Paul

(Transport)

J. J. Hyland, Commander, USN
J. H. Conyne, Lieutenant, USN
N. O. Wynkoop, Lieutenant, USN
J. C. Walle, Lieutenant (jg), USNRF
C. Keenan, Ensign (T), USN
J. J. McCarey, P. A. Surgeon, NNV
B. R. Lyon, Asst. Surgeon, USN
A. F. Jeffrey, 'Boatswain (T), USN
J. Bona, Gunner (T), USN
J. C. Stephenson, Gunner (T), USN
W. G. McIntyre, Carpenter (T), USN
A. L. Crowder, Pharmacist (T), USN

U. S. S. San Diego

(Cruiser)

H. H. Christy, Captain, USN
C. B. Price, Commander, USN
R. C. McFall. Lieut. Commander, USN
J. S. McCain, Lieutenant, USN
R. R. Zivnuska, P. A. Paymaster, USN
R. J. Carstarphen, Lieutenant, USN
C. E. Hoard, Lieutenant (jg), USN
F. G. Kutz. Ensign, USN
F. S. Irby. Ensign. USN
A. G. Reaves, Ensign, USN
P. F. Shortridge, Ensign. USN
W. R. Buchner, Ensign, USN
J. C. Collins, Ensign. USN
W. Henderson. Ensign. USNRF
D. M. Stewart, Lieut. Comdr., NNV
A. B. Adams, Lieutenant, NNV
C. T. Wallace. P. A. Surgeon, NNV
L W. Parson, Asst. Surgeon, NNV
H. W. Lewis, Lieutenant, USNRF
T. A. Stetson, Lieutenant (jg), USNRF
W. H. Melseme, Lieut, (jg), USNRF
C. M. Johnston, Boatswain. USNRF
J. C. Short, Chaplain, USNIIF
C. H. Uznay, Ensign, USNRF
K. E. Hintze, Lieutenant (jg), USNRF
C. E. Kuter, Ensign, USNRF
H. H. Searles, Asst. Surgeon, USNRF
D. W. Loomis, Ensign, USNRF
J, D. Murray, Ensign, USNRF
R. D. Joldersma, Asst. Surg., USNRF
G. A. Browne, Lieutenant, NNV
A. C. Kidd, Ensign, USN

G. A. Beall, Lieutenant, USN
R. C. Jones, Lieutenant, USNRF
B. G. Barthalow, Lieut. Comdr., USN
E. O. J. Eytinge, P.A. Surgeon, USN
L. L. Babbitt, Lieutenant (jg), USN
S. G. Meyer, Gunner, USN
D. J. Burke, Gunner (Radio), USN
J. D. Gagan, Pay Clerk, USN
R. J. Monteith, Pay Clerk, USN

Relief Officers in Order of Reporting on Board

G. S. Gillispie, Lieutenant (jg), USN
S. M. Akerstrom, Dental Surgeon, USN
L. R. Holm, Ensign, NNV
D, Easdale, Carpenter, USN
A. Henderson, Boatswain (T), USN
J. B. Cadenbach, Boatswain (T), USN
J. P. Hildman, Gunner (T), USN
J. B. Dofflemeyer, Gunner (T), USNRF
D. J. Burke, Gunner (T), USNRF
W. A. Zellar, Machinist (T), USN
F, B. Devlin, Ensign, NNV
G. Watts, Captain, USMC
D. Kenyon, Ensign, USMC
V. G. Greiff. Lieutenant (jg), USNRF
G. Bradford, Lieut. Commander, USN
J. H. Russell. Ensign (T), USN
A. V. Jannotta, Ensign (T), USN
R. F. Sheehan, P. A. Surgeon, USN
G. F. Adams, Captain, USN
G. W. Reihle, Ensign, USNRF
T. P. Lovelace, Lieutenant (jg), USNRF
V. E. Harkness, Lieutenant (jg), USNRF
W. B. Cowan, Ensign, USNRF
J. B. Dewar, Ensign, USNRF
V. W. Hickman, Ensign, USNRF
W. Knight, Ensign, USNRF
J. B. Malone, Ensign, USNRF
R. B. McCauley, Ensign, USNRF
A. H. Gerbig, Act. Pay Clerk, USN
J. J. Lucas. Ensign (T) (G), USN
H. C. Petterson, Asst. Surgeon, USN
C. E. Egeler, Lieutenant (jg), USNRF
D. A. Owens, Captain, USN
H. C. Hemingway, 2nd Lieut., USCG
C. J. Bright, Lieutenant, USN
D. Holbrook, Ensign, USNRF
P. G. Kent, Ensign, USNRF
W. J. Murray, Ensign. USNRF
J. P. O'Riorden, Ensign, USNRF
C. O. Osborne, Ensign, USNRF

C. E. Peters, Ensign, USNRF
C. H. Seils, Ensign, USNRF
J. Stewart, Jr., Ensign, USNRF
H. H. Taylor, Ensign, USNRF
H. D. Whitcomb, Ensign, USNRF

U. S. S. Seattle

*Flagship of Vice Admiral Albert Gleaves, USN
(Cruiser)*

De Witt Blamer, Captain, USN
R. Drace White, Lieut. Comdr., USN
Austin S. Kibbee, Lieut. Comdr., USN
Charles C. Gill, Lieutenant, USN
Philip H. Hammond, Lieutenant, USN
Henry E. Parsons, Lieutenant, USN
John L. Callan, Lieutenant, USNRF
Robertson J. Weeks, Lieut, (jg), USN
William H. Burtis, Lieut, (jg), USN
L. Lee Babbitt, Lieut. O'g), USN
Robert M. Farrar, Ensign, USN
George H. Keller, Ensign, USN
Frank L. Worden, Ensign, USN
Thomas L. Hendley, Ensign, USN
John H. Forshew, Ensign, USN
James P. Conover, Ensign, USN
Constantine N. Perkins, Ensign, USN
Chester M. Holton, Ensign, USN
Robert A. Haynie, Ensign, NNV
Alfred L. Clifton, P.A. Surgeon, USN
Philip J. Murphy. Asst. Surg.. USNRF
Alonzo G. Hearne, P. A. Paym., USN
Richard H. Tebbs, Jr., Captain. USMC
August Wohltman, Chief Bosn., USN
Joseph M. Gately, Gunner (Ord.), USN
Thomas Flynn, Gunner (Elec), USN
Arthur Boquett, Gunner (Radio), USN
William W. Holton, Machinist, USN
John A. Silva, Machinist, USN
Abram A. Broughton, Machinist, USN
Goldsboro Sessions, Carpenter, USN
Clarence A. Miley, Pay Clerk, USN
Charles W. Pearles, Gunner (El.), USN
Frank S. Miller, Gunner (Radio), USN
Harold R. Lehmann, Act. Pay Clk, USN

*Relief Officers in Order of Reporting on
Board*

Joseph Baer, Lieutenant, USN
Andrew McCreey, Ensign, USNRF
John Black, Ensign, USNRF
John W. Collier, Boatswain, USN

Ewell K. Jett, Gunner (Radio), USN
Homer H. Simons, Gunner (Elec.) USN
G. Irwin Kohlmeir, Dental Surg., USN
C. Maple, Asst. Paymaster, USNRF
Edward H. Sparkman, P.A. Surg., NNV
L. F. Randall, Act. Pay Clerk, USN
John S, Waters, Lieut, (jg), USN
De Courcy Fales, Ensign, NNV
J. Sherlock Archibald, Ensign, NNV
Carl S. Ziesel, Dental Surgeon, USN
John S. Putnam, Act. Chaplain, USN
L. Clowney, Ensign, NNV
C. R. Jacobson, Ensign, NNV
L. H. Van Syckle, Ensign, NNV
H. Duggan, Asst. Surgeon, USNRF
Francis H. Stone, Ensign, USNRF
William H. Long, Ensign, USNRF
Walcott Blair. Ensign, USNRF
Edward H. Wardell, Ensign, USNRF
Robert C. Huneke, Boatswain, USN
Spencer Nickols, Ensign, NNV
Duncan P. Howser, Ensign, USNRF
Nelson J. Leonard, Lieutenant, USN
Butler Y. Rhodes, Lieut. Comdr., USN
Frederick A. Savage, Lieut. Comdr., USN
Alfred E. Stulb, Ensign (T), USN
Harry W. Hosford, Lieutenant, USN
Ralph S. Rankin, Lieut, (jg), USNRF
F. H. Geer, Ensign, USNRF
W. K. Stevenson, Boatswain (T), USN
J. R. Y. Blakely, Captain, USN
E. H. Friedman, Ensign, USN
Dean D. Francis, Ensign, USN
R. F. Taylor, Pay Clerk, USNRF
Jesse B. Oldendorf, Lieutenant, USN
H. L. Willoughby, Jr. (jg), USNRF
Howard F. Devlin, Ensign, USNRF
John F. Sheldon, Ensign, USNRF
Joseph W. Rixey, Ensign, USNRF
Henty A. Williams, Ensign, USNRF
Hammond C. Bowman, Ensign, USNRF
Jesse J. Borschman, Cadet, USNRF
Lucius W. Smith, Cadet, USNRF
Lindsley F. Kimball, Cadet, USNRF
William L. Culbertson, Comdr., USN
Elbert C. Isom, Ensign, USNRF
John S. Waterman, Jr., Ensign, USNRF
Melvin L. Southwick, Ensign, USNRF
Arthur E. Navlet, Ensign, USNRF
Eric G. Hoylman, Dental Surg., USN
Louis F. Peifer, 2nd Lieutenant, USMC
Charles Maiden, Machinist (T), USN

Dion W. Taylor, Carpenter, USN
Henry M. Cowardin, Ensign, USNRF
John Gordon, Gunner (T), USN
John Costello, Gunner (T), USN
Ralph U. Clark, Gunner (T), USN
Albert J. Claussen, Machinist (T), USN
Stewart L. Johnson, Mach. (T), USN
G. V. Vail, Lieut, (jg) (DC), USNRF

U. S. S. Sialia

Flagship of Rear-Admiral H. P. Jones, USN

Samuel W. King, Lieutenant, USN
W. Charles M. Clark, Lieut, (jg), NNV
George S. Silsbee, Ensign, USNRF
Charles F. Fretz, Ensign, USNRF
Roy T. Bookmeyer, Ensign, USNRF
Worth B. Beacham, Ensign, USNRF

Relief Officers in Order of Reporting on Board

W. F. Burruss, Ensign, USNRF
Ray W. Clark, Lieutenant, USNRF
W. H. Theisen, Gunner (R), USN
O. P. Shattuck, Lieutenant (jg), USN
P. D. Reynolds, Ensign, USNRF

U. S. S. Siboney

(Transport)

A. T. Graham, Commander, USN
IL C. Train, Lieut. Commander, USN
E. P. A. Simpson, Lieutenant, USN
N. Withers, Lieutenant, USN
C. W. Henckler, Ensign (T), USN
T. (t. Summers, Asst. Surgeon, USN
C. R. Murray, Asst. Paymaster, USN
W. H. Misch, Pay Clerk, USN
A. F. Goodrich, Boatswain, U'SN
H. M. Scidschlag Gunner (Ord.), USN
J. IL Conroy, Machinist, USN
J. C'. Kreiger, Machinist, USN
O. E. Whilden, Machinist, USN
M. W. Jones, Carpenter, USN
O. D. Sipe, Pharmacist, USN
C. E. Kreml, Pharmacist, USN
A. J. Murietta, P.A. Surgeon, NN '
W. T. Burgess, Ensign, NNV
J. D. McLeod, Lieutenant, USNRF
A. Daunt, Lieutenant (jg), USNRF
G. R. Griffith, Lieutenant (jg). USNRF

J. Roth, Ensign, USNRF
C. C. Fales, Ensign, USNRF
C. Castelloe, Asst. Surgeon, USNRF
H. A. McKay, Lieutenant (jg), USNRF
M. R. Coward, Ensign, USNRF
J. J. Finnegan, Ensign, USNRF
J. I. Eckford, Asst. Paymaster, USNRF
R. A. La Bine, Asst. Surgeon, USNRF .
W. N. Updegraff. Ensign, USN
G. T. Palmer, Ensign. USNRF
T. E. D. Veeder, Jr., Ensign, USN
M. H. Pingree, Ensign, USNRF
F. H. Preti, Ensign, USNRF
T. A. Printon. Ensign, USNRF
A. L. Robinson, Ensign, USNRF
W. J. Roberts, Ensign, USNRF
T. J. Reynolds, Ensign, USNRF
M. Y. Parker, Ensign, USNRF

Relief Officers in Order of Reporting on Board

R. Morris, Commander, USN
W. F. Besse, Lieutenant. USNRF
R. B. Sawyer, Lieutenant, USNRF
J. Gibbons. Lieutenant (jg), USNRF
C. Reimann, Gunner (O), USN
P. B. Marzoni, Lieutenant, USNRF
W. S. Gable, Ensign, USNRF
Jos. A. O'Donnell, Chf. Gunner, USNRF
G. B. Martinson, Boatswain, USN
J. A. A. Somblom, Boatswain, USN
W. H. Sheffield, Ensign (T), USN
H. D. Scott, Ensign (T). USN
R. G. Seger, Ensign (T), USN
C. S. Williams, Lieutenant, USN
John D. Kennedy, Ensign, USNRF (Temporary duty)
Martin J. Jukich, Ensign, USNRF (Temporary duty)
L. P. Kane, Ensign, USNRF (Temporary duty)
H. F. Parks, Ensign, USNRF (Temporary duty)
H. A. Seran, Lieut. Comdr.. USNRF
G. E. Robertson, Ensign, USNRF
F. M. Scribner, Ensign (T), USN
S. L. Jeffrey, Lieutenant (MC), USNRF
J. N. Mosher, Machinist, USNRF
H. B. Delcamp, Machinist, USNRF
G. V. Smith, Machinist, USNRF
L. H. Lancaster, Machinist, USNRF
Franklin G. Wright, Lieut., USNRF
George W. Calbeck, Ensign, USNRF

U. S. S. Sierra

(Transport)

J. D. Willson, Commander, USN
L. S. Stewart, Lieut. Comdr., USN
A. E. Younie, Lieut. Comdr. (MC), USN
F. M. Cook, Lieutenant, USNRF
C. H. Hurley, Lieutenant, USNRF
C. T. Anderson, Lieut, (jg), USNRF
S. N. Blossom, Lieutenant (jg), USNRF
C. D. Draper, Lieutenant (jg), USNRF
J. W. Dunn, Lieutenant (jg), USNRF
F. S. Durden, Lieutenant (jg), USNRF
C. C. Makin, Lieutenant (jg), USNRF
W. M. Mullin, Lieutenant (jg), USNRF
J. W. Rowe, Lieutenant (jg), USNRF
C. A. Wood, Ensign (T), USN
M. H. Abells, Ensign, USNRF
W. Dickey, Ensign, USNRF
A. Hall, Ensign, USNRF
H. E. La Mertha, Ensign, USNRF
J. Metcalf, Ensign, USNRF
F. L. Fichard, Ensign, USNRF
H. E. Scott, Ensign, USNRF
M. M. Braff, Lieutenant (MC), USN
W. A. Bacon, Lieutenant (MC). USN
G. Rembert, Lieutenant (PC), USN
M. Tuthill, Ensign (PC), USNRF
M. Ballord, Gunner (T), USN
P. H. Scribante, Bosn. (T), USN
L. C. Wishard, Gunner (T), USN
E. J. Beynton, Gunner, USNRF
R. C. Hiby, Machinist, USNRF
F. H. Thames, Carpenter (T). USN
T. E. Wiggins, Pharmacist (T). USN
W. O. Wood, Act. Pay Clerk (T), USN

U. S. S. South Dakota

(Cruiser)

Lucius A. Bostwick, Captain, USN
Merlin G. Cook, Lieut. Comdr., USN
Lindsay H. Lacy, Lieut. Comdr., USN
George H. Blair, Lieut. Comdr., NNV
Edmund D. Almy, Lieutenant, USN
Herbert A. Jones, Lieutenant, USN
Tracy L. McCauley, Lieutenant, USN
Wedell Foss, Lieutenant, NNV
George H. Jett, Lieutenant, NNV
Harold P. Parmelee, Lieut, (jg), USN
William A. Corn, Lieut, (jg), USN

Frank P. Thomas, Lieut, (jg), USN
Julius M. Moss, Lieut, (jg), USN
F. R. Dodge, Ensign, NNV
Frederick L. Douthit, Ensign, USN
Jack S. Phillips, Ensign, USN
W. K. Phillips (on leave). Ensign, USN
Edw. H. LeTourneau, Ensign, NNV
Henry R. Wakeman (EDO), Ensign, NNV
Russell A. Mackey, Ensign, NNV
Wm. R. McAdam, Ensign, NNV
Ernest C. May, Ensign, USNRF
Frederick T. Montgomery, Act. Ensign, USN
Adolph Peterson, Ensign, USN
Oscar Benson, Ensign, USN
Stephen A. Loftus, Ensign, USN
Robert J. Kingsmill, Ensign, USN
George R. Blauvelt, Ensign, USN
Carleton I. Wood, Asst. Surgeon, USN
Ben L. Norden, Asst. Surgeon, NNV
Ralph W. Swearingen, Asst. Pay., USN
Samuel H. Kiiowles, Asst. Paymaster (T), USN
Geo. G. Schweizer, Asst. Pay. (T), USN
Thomas P. Kane, Boatswain (T), USN
Coenraad Lichtendahl, Boatswain (T), USN
Loar Mansbach, Gunner (E) (T), USN
Louis M. Palmer, Gunner (T), USN
Arthur F. Armstrong, Gunner (T), USN
G. H. Toepfer, Carpenter (T), USN
R. Anderson, Gunner (T) (R), USN
O. L. de Vasconcelles (Secunde Tenente, Brazilian Navy)

Relief Officers in Order of Reporting on Board

Archibald Young, Captain, USMC
Milton W. Vedder, 2nd Lieut., USMC
Geo. F. Kelly, Asst. Surgeon, USNRF
Clarence B. Archer, Act. Pay Clk., USN
Guy E. Thornton, Gunner (T), USN
Charles H. Gillilan, Pay Clk. (T). USN
Donald Butter, Lieutenant (jg), USN
W. T. Dabney, Lieutenant (jg), USNRF
C. D. Gibbs. Lieutenant (jg), USNRF
R. E. Kinkead, Lieutenant (jg), USNRF
T. Marceau, Ensign (T), USN
F. M. Smith, Ensign (T), USN
C. E. Miller, Ensign (T), USN
Roy Childs, Gunner (T), USN
Grover Williams, Gunner (T), USN
Herbert C. Conner, Machinist (T), USN
William T. Evans,_ Machinist (T), USN

Francis W. Orpin (EDO), Ensign, USNRF
G. H. Frederick, Ensign (T), USN
A. M. O. Wood, Asst. Surgeon, USNRF
George Crofton, Lieut. (jg) (T), USN
L. C. Harris, Acting Chaplain, USN
William Condon, Gunner (T), USN
William H. Glasper, Gunner (T), USN
Harry V. Kelly, Boatswain (T). USN
John J. Solosky, Pay Clerk (T), USN
Roy Chllds, Ensign (T), USN
D. W. Hand, Ensign, USN
E. E. Herrmann, Ensign, USN
H. P. Kirby, Midshipman, USN
L. C. Lawbaugh, Midshipman, USN
W. I. Leahy, Midshipman, USN
C. V. Lee, Midshipman, USN
E. Lewis, Midshipman, USN
J. A. Lusk, Midshipman, USN
H. D. Lyttle, Midshipman, USN
E. D. McEathron, Midshipman, USN
D. A. Hughes, Midshipman, USN
Harold W. Gamble, Dental Surgeon, USNRF
John F. Donelson, Lieutenant, USN
Leland S. Swindler, 2nd Lieut., USMC
John M. Luby, Captain, USN
Turner F. Caldwell, Commander, USN
George B. Keester, Lieut. Comdr., USN
Elliott F. Pettigrew, Ensign (T), USN
Linn D. Shipman, Ensign (T), USN
Albert P. Rumsey, Ensign (T), USN
Omar T. Pfeiffer, Captain. USMC
George E. Henning, Boatswain (T), USN
John L. Matthews, Ensign (T), USN
Ralph S. Maughan. Ensign (T), USN
Walter B. Holder, Ensign (T), USN
Edward D. Magill, Ensign (T), USN
John A. Meyer, Ensign (T), USN
Mackey C. Saylor, Gunner (T), USN

U. S. S. Susquehanna
(Transport)

Z. H. Madison, Commander, USN
A. B. Reed, Lieut. Commander, USN
R. S. Chew, Jr., P.A, Paymaster, USN
M. Hudson, Lieutenant (jg), USN
L. B. Scott, Ensign, USN
C. H. Hosung, Ensign, USN
E. H. Faro, Ensign, USNRF
O. T. Purcell, Ensign, USN
W. M. Snell, Ensign, USNRF
L. W. Gumz, Boatswain, USN

R. S. Savin, Gunner, USN
G. O. Farnsworth, Gunner, USN
H. R. Newby, Carpenter, USN
H. W. Niels. Ensign, USNRF
W. T. Oppenheimer, Asst. Surg., USN
L. A. Puckett, Act. Pay Clerk, USN
J. E. Malcolmson, Asst. Surgeon, USN
R. E. Morton, Asst. Paymaster, USN

Relief Officers in Order of Reporting on Board

B. D. Schmidt, Asst. Paymaster, USN
W. E. Davis, Asst. Paymaster, USN
G. G. Holliday, P. A. Surg.. USNRF
R. G. Avery, Lieutenant (jg), USNRF
C. M. Cain, Machinist (T). USN
A.ffl. George, Lieutenant, USNRF
H.H. Williamson, Pharmacist, USNRF
A. O. Mundale, Boatswain (T), USN
W. B. Anderson, Gunner, USN
C. Wright. Machinist (T), USN
M. A. Beach, Carpenter (T), USN
H. M. Home, Lieutenant, USN
F. S. Williams, Lieutenant, USNRF
P. S. Stewart, Lieutenant (jg), USNRF
W. P. Ames, Ensign, USNRF
J. D. Herbert, Ensign, USNRF
G. T. Boone, Ensign, USNRF
J. G. Lewis, Ensign, USNRF
R. A. Gilbert, Asst. Surgeon, USN
C. C. Wheeler, Chap. Lt. (jg), USNRF
J. W. Whitney, Lieut. (jg), USN
A. C. Schroeder, Lieut. (jg), USNRF
S. H. Packer, Ensign, USNRF
H. G. W. Parmele, Ensign, USNRF
W. L. Martin, Asst. Surgeon, USN
F. L. Buckley, Asst. Paym., USNRF
R. P. Morse, Dental Surgeon, NNV
E. R. Perkins, Act. Pay Clerk (T), USN
V. B. Felitto, Gunner, USNRF
J. C. M. Small, Ensign (T), USN
G. F. Rieman, Ensign (T), USN
D. P. Robinson, Ensign (T), USN
E. L. Stites. Ensign (T), USN
J. F. Sullivan, Ensign (T), USN
H. S. Warren, Ensign (T), USN
G. M. West, Ensign (T), USN
J. L. Freese, Ensign (T). USN
W. T. Davidson, Dental Surgeon. USN
C. Castelloe, Asst. Surgeon, USN
W. A. Busse, Ensign, USNRF

F. E. Graves, Ensign, USNRF
T. W. Greenland, Ensign, USNRF
A, Jablons, Asst. Surgeon, USN
H. H. Fenskov. Boatswain (T). USN
E. H. Vanderbeck, Gunner (T). USN
C. C. Fry, Machinist (T). USN
M. A. Beach, Carpenter (T), USN
M. P.Hanlon, Lieut, (jg), USNRF (MC)
J. McNaught, Lieut, (jg), USNRF
J. Shaw, Act. Pay Clerk (T), USN
W. P. Mull, Lieutenant (MC), USN
W. C. Carroll, Lieut. (DC) (jg), USNRF

U. S. S. Tacoma
(Cruiser)

Powers Symington, Commander, USN
G. M. Cook, Lieutenant (jg), USN
J. Garnett, Lieutenant (jg), USN
E. Buckmaster, Lieutenant (jg), USN
H. H. Crow, Lieutenant (jg), NNV
C. A. Baker, Ensign, USN
P. R. Glutting, Ensign, USN
T. R. Solberg, Ensign, USN
J. Fife, Midshipman, USN
G. Rowe, Midshipman, USN
P. K. Fischler, Midshipman, USN
R. M. Hayes, A. Surgeon, USNRF
W. A. Carey, P. A. Paymaster, NNV
J. H. Ranch. Asst. Paym. (T), USN

Relief Officers in Order of Reporting on Board

W. B. Jones, Act. Pay Clerk (T), USN
H. M. Corse, Lieut, (jg), USNRF
A. W. Ford, Ensign, USNRF
S. Dillon, Ensign, USNRF
R. H. Cobb, Ensign, USNRF
W. D. I.IacDougall, Captain, USN
R. L. Rowan, Asst. Paymaster, NNV
F. J. Sexton, 2nd Lieutenant, USCG
Duncan M. Wood, Captain, USN
D. Scott, Ensign (T), USN
P. Lee, Ensign (T), USN
N. D. Weir, Ensign (T), USN
F. Pettit, Jr., Ensign. USN
R. B. Netting, Ensign, USN
R. F. Nelson, Ensign, USN
H. A. Ellis, Lieutenant, USN
C. W. Jagger, Ensign (D), USNRF
H. W. Proom, Ensign (D), USNRF

R. J. Costigan, Ensign (D), USNRF
J. L. Kahle, Ensign (D), USNRF
J. S. Dean, Ensign (D), USNRF
W. B. Leake, Asst. Surgeon, USNRF

U. S. S. Tenadores
(Transport)
Army Transport — Officers Serving Prior to April 17, 1918

W. R. Sexton, Captain, USN
Robert Henderson, Commander, USN
B. Y. Rhodes, Lieut. Commander, USN
R. M. Comfort, Lieut. Comdr., USN

(Taken over by Cruiser and Transport Force in April, 1918)

Officers Serving After April 17, 1918

John D. Wainwright, Commander, USN
W. T. Mallison, Lieut. Comdr., USN
H. J. Grassie, Lieutenant, USN
C. J. Brobeck, P. A. Surgeon, USNRF
F. E. O'Brien, Asst. Surgeon, USNRF
R, Pelliconi, Lieutenant, USNRF
R. C. DouU, Lieutenant, USNRF
L. W. Busbey, Lieutenant (jg), USNRF
R. A. Gawley, Lieutenant (jg), USNRF
J. N. Nichols. Asst. Surgeon. USNRF
W. S. Rhoades, Asst. Paymaster, USN
J. S. Collins, Asst. Paymaster, USNRF
I. W. Jenkins, Lieutenant (jg), USNRF
J. Marcussen, Lieutenant (jg), USNRF
W. I. Green, Lieutenant (jg), USNRF
O. Eversen, Lieutenant (jg), USNRF
George Merkle, Ensign, USNRF
W. R. Kellam, Ensign, USNRF
N. Jones, Ensign, USNRF
E. A. Bell, Ensign, USNRF
R. R. Averill, Ensign, USNRF
M. J. O'Donnell, Ensign, USNRF
W. D. Chambers, Ensign, USNRF
F. F. Law, War. Machinist, USN
W. W. Hobelman, Carpenter, USN
L. S. Tichener, Electric Gunner, USN
T. McGann. Gunner, USN
C. H. Spearman, Pharmacist, USN
J. E. Wood, Acting Pay Clerk, USN

Relief Officers in Order of Reporting on Board

H. R. Thurber, Ensign, USN

R. B. Tuggle, Ensign, USN
P. M. Clark, Ensign, USNRF
L. C. Claff, Ensign, USNRF
D. Crowell. Ensign. USNRF
J. S. Carpenter, Ensign, USNRF
W. J. Ilazelwood. Lieut, (jg), USNRF
A. G. Reaves, Lieutenant, USN
C. T. Sprout, Ensign, USNRF
F. H. Stecher, Ensign, USNRF
C. R. Steames, Ensign, USNRF
C. C. Stevenson, Ensign, USNRF
E. G. Reynolds, Ensign, USNRF
Louis J. Connelly, Captain, USN
O. T. Miller, Lieutenant, USN
J. Harder, Lieutenant. USN
F. J. Bailey, Lieutenant, USNRF
J. S. Rosenthal, Lieut. (MC), USNRF
E. B. Rinker, Lieutenant (jg), USN
W. M. Connelly, Lieut, (jg), USNRF
F. A. Mulcahy. Ensign, USNRF
H. Schmitz, Ensign, USNRF
J. T. Leonard, Ensign, USNRF
H. K. Hutchins, Ensign, USNRF

U, S. S. Von Steuben

(Cruiser)

Stanford E. Moses, Commander, USN
Charles II. Bullock, Lieutenant, USN
Fred T. Berry, Lieutenant, USN
Joseph M. Mitcheson, Lieut., USNRF
Joseph F. Kaney, Lieutenant, USNRF
P. C. Cornelius, Lieut, (jg), USNRF
Eugene Ames, Ensign, USNRF
E. R. Cassidy, Ensign, USNRF
F. H. Lemly, P. A. Paymaster, USNRF
A. E. Biddinger, P. A. Surgeon, NNV
J. G. Harvey, P. A. Surgeon, NNV
D. Dowling, Chief Boatswain, USN
L. Haase, Chief Carpenter, USN
E. Fisher, Gunner, USN
E. A. Healy, Machinist, USN
F. W. Shepard, Machinist, USN

Relief Officers in Order of Reporting on Board

P. W. Hathaway, Pay Clerk, USN
Fred J. Butterfield, Lieut, (jg), USNRF
James L. Fisher, Ensign, USN
John M. Haines, Ensign, USN
Richard F. Norvell, Ensign, USNRF
Augustus H. Wordell, Ensign, USNRF

Eric P. Teschner, Machinist, USN
Emil A. Lichtenstein, Lieutenant, USN
Andrew B. Williams, Pay Clk., USNRF
Niles A. Bolin, Lieut, (jg) (D), USNRF
Chas. M. Mundie, Lieut, (jg), USNRF
C. D. Bennett, Lieut, (jg), USNRF
James C. McDermott, Ensign (T), (G), USN
Nathan K. Bassarear, Ensign (D), USNRF
E. T. Comins, Ensign, USNRF
Yates Stirling, Captain, USN
H. E. Rideout, 1st Lieutenant, USCG
Edgar O. Galyon, Lieut, (jg), USNRF
V. E. Babington, Asst. Surgeon, USNRF
P. F. F. Wangerin, Ensign, USNRF
Jesse F. Spink, Ensign, USNRF
George H. Sipp, Gunner (T), USN
James Sanders, Carpenter (T), USN
H. J. Gillen, Act. Pay Clerk, USNRF
A. B. Weinberger, Lieut, (jg), USNRF
D. G. Schmitz, Ensign (T), USN
H. Hilton-Green, Ensign (T), USN
E. C. Haaren, Ensign (T), USN
K. W. Mayo, Dental Surgeon, USNRF
D. M. Waesche, Ensign, USNRF
M. Palmer, Ensign, USNRF
H. D. Grinnell, Ensign, USNRF
H. Harvey, Ensign, USNRF
R. L. Tegart, Asst. Paym., USNRF
R. K. Bonsteel, Asst. Paym., USNRF
E. Stohan, Boatswain (T), USN
C. Shilasky, Boatswain, USNRF
S. D. Morgan, Machinist, USNRF
J. J. Comas, Machinist, USNRF
N. Jensen, Machinist (T), USN
C. Fairman, Pay Clerk (T), USN
H. Brumberger, Gunner (T), USN
R. M. Ihrig, Ensign, USN
R. E. Jennings, Ensign, USN
Frank Pardee, Jr., Ensign, USNRF
W. B, McCormick, Ensign, USNRF
R. Rowland, Ensign, USNRF
W. M. Evans, Gunner (E), USNRF
W. J. Dragon, Machinist (T), USN
F. A. Brannen, Lieut, (jg), USNRF
W. E. Davis, Asst. Paymaster, USN
Mason Scudder, Ensign, USNRF
W. W. Ryan, Ensign, USNRF
J. L. McHenry, Ensign, USNRF
J. E. Pilkington. Ensign, USNRF
S. A. Mann. Pay Clerk, USNRF
C. R. Miller, Captain, USN

J. W. Wilcox, Commander, USN
Chas. Blount, Ensign, USNRF
J. Baker, Ensign, USNRF
B. A. Larsen, P. A. Surgeon, USNRF
W. P. Gaddis, Lieut. Comdr., USN
H. Langworthy, Lieutenant, USNRF
M. G. Tucker, Ensign (T), USN
E. S. Underhill, Ensign (T), USN
H. Schaetzle, Ensign (T), USN
J. F. Roth, Ensign (T), USN
W. H. Van Wart, Ensign (T), USN
John B. Ford, Ensign (T), USN
Fred S. Treat, Ensign (T), USN
Max von Schrader, Ensign (T), USN
W. Guerry, Lieut, (jg) (PC), USN

U. S. S. Wilhelmina
(Transport)

W. T. Tarrant, Commander, USN
J. B. Gilmer, Lieut. Commander, USN
J. W. Jory, Lieut. Commander, USNRF
W, I. Causey, Lieutenant, USN
D. Caldwell, Lieut, (jg), USNRF
G. A. Berndtson, Lieut, (jg), USNRF
I. T. Fahlberg, Lieut, (jg), USNRF
W. Van Houton, Ensign, USNRF
G. W. Walton, Ensign, USNRF
F. A. Willard, Ensign, USNRF
E. F. Baldwin. Ensign, USNRF
J. Young, Lieutenant, USNRF
J. G. Hutchinson, Lieut, (jg), USNRF
F. J. Rogers, Lieut, (jg), USNRF
E. P. Herney, Ensign, USNRF
J. O. Crom, Machinist, USN
H. L. Brown, Surgeon, USN
C. D. Shannon, Asst. Surgeon, USN
C. M. McKee, Asst. Surgeon, USNRF
J. F. LaSalle, Pharmacist, USN
A. R. Schofield, P. A. Paymaster, USN
C. D. Everingham, Asst. Paym., USNRF
F. C. Welch, Asst. Paymaster. USNRF
W. F. Brown, Pay Clerk, USN
D. J. Lewis, Pay Clerk, USNRF
H. P. Gleason, Boatswain, USN
G. B. Dahlman, Gunner, USN
J. F. Shea, Gunner, USN
W. Stoudt, Carpenter, USN

*Relief Officers in Order of Reporting on
Board*

John Grady, Commander, USN

E. F. Manning, Ensign, USN
F. J. Scheufele, Ensign, USNRF
L. W. Smith, Ensign, USNRF
A. A. Sayres, Ensign, USNRF
F. C. Seymour, Ensign, USNRF
P. S. Sampson, Ensign, USNRF
G. B. Ruggles, Ensign, USNRF
A. H. Schow, Ensign, USNRF
T. W. Salmon, Ensign, USNRF
C. Foose, Lieutenant (jg), USNRF
R. J. Routledge, Lieut, (jg), USNRF
H. L. Sweetser, Ensign, USNRF
D. W. Ladd, Ensign, USNRF
J. A. Lunn, Ensign, USNRF
N. P. Patterson, Ensign, USNRF
J. C. Paden, Lieutenant (MC), USN
J. P. Worsham, Pay Clerk, USNRF
G. W. Clark, Lieutenant, USN
R. J. Crocker, Lieutenant (jg), USNRF
I. Jacobson, Lieutenant (jg), USNRF
H. F. Reid, Ensign (T), USN
J. R. Witbeck, Ensign (T), USN
W. J. Wolf, Ensign (T), USN
F. R. Uhlig, Ensign (T), USN
W. M. Toomey, Ensign (T), USN
A. S. Whitehead, Lieutenant, USNRF
D. M. Yoder, Ensign (T), USN

U. S. S. Yorktown
(Cruiser)

William H. Allen, Commander, USN
David C. Guest, Lieut. Comdr., USNRF
Frank L. Lowe, Lieutenant, USN
Walker P. Rodman, Lieut. (T) (jg), USN
Clyde Morrison, Lieut. (T) (jg), USN
Walter E. Torrey, Lieut, (jg), USNRF
Raymond M. Bright, Lieut. (PC), USN
Thomas F. Long, Lieut. (MC), USN
Daniel F. Black, Ensign, USNRF
Linton H. Smith, Ensign, USNRF
Henry L. Bray, Ensign, USNRF
LeRoy L. Carver, Ensign, USNRF
Paul C. Noble, Ensign, USNRF
Chauncey G. Ollinger, Pay Clerk, Act., USN
Charles F. Clark, Boatswain (T), USN
Carl Herrick, Pay Clerk (T), USN

U. S. S. Zeelandia
(Transport)

Robert Henderson, Commander, USN

J. J. McCracken, Lieut Comdr., USN
W. J. Blake, Lieutenant, USNRF
C. W. Weitzel, Lieutenant, USN
J. W. Kirschner, Lieutenant, USNRF
Elliot Ranney, Asst. Paymaster, Lieut., USNRF
R. A. McDonnell, Supply Off. Lt. (jg), USNRF
William T. White, Lt.(jg) (D), USNRF
Victor J. Noel, Lieut, (jg) (D, USNRF
Harry D. Chemnitz, Lieut, (jg) (D), USNRF
E. V. Ferrandini, Lt. (jg) (D). USNRF
Ausey H. Robnett, Surgeon, USN
Ed. A. Mullen, Asst. Surgeon, USN
Jno. A. B. Sinclair, Act. Asst. Surg., USN
B. E. Munroe. Ensign (D), USNRF
S. F. Houston, Ensign (D), USNRF
R. E. Lindorff, Ensign (D), USNRF
G. W. E. Mikkelson, Ensign (E), USNRF
George P. Hynes, Ensign (E), USNRF
E. F. Alward, Ensign (D), USNRF
Leslie R. Bristow, Ensign (D) USNRF
Dwight C. Ely, Ensign (D), USNRF
Henry T. Mitchell, Lieut, (jg) (E), USN
F. W. Atherton, Boatswain, USN
C. R. Brown, Gunner (E), USN
W. H. Buchanan, Carpenter, USN
F. Carter, Machinist, USN
Edward P. Lapp, Pay Clerk, USN
Louis Lindenmayer, Pay Clerk, USN
Charles F. Whitmore, Pharmacist, USN
Thomas M. Diegnan, Gunner (E), USN
R. N. Rindernecht, Ensign (D), USNRF

Relief Officers in Order of Reporting on Board

F. H. Babcock, Asst. Paymaster, Ensign, US-NRF
John A. Salb, Asst. Surgeon, USNRF
James F. McGrath, Dental Surg., USN
P. M. Woodwell, Ensign (T) (D), USN
John Walter Young, Ensign (D), USNRF
Maurice A. Malandain, Lieut, (jg) (E), USNRF
Oliver H. Clark, Ensign (E), USNRF
David Lyons, Commander, USN
W. C. Morhoff, Lieutenant, USNRF
Chas. A. Wilson, Ensign, USNRF
Frank H. Smith, Ensign, USNRF
R. B. Scharman, Ensign, USNRF
Harry C. Evans, Ensign, USNRF
W. L. Radcliff, Ensign, USNRF
J. S. Sutton, Ensign, USNRF

H. E. Clark, Boatswain, USN
E, A. Lichtenstein, Lieut. Comdr., USN
G, G. Meade, Lieutenant, USNRF
R. A. Fox, Ensign, USNRF
E. Daskam, Ensign, USNRF
R. F. Tillman, Ensign (T), USN
P. H. Weinberg, Ensign (T), USN
R. McK. Stover, Ensign (T), USN
P. A. Thompson, Ensign (T), USN
C. C. Reynolds, Ensign (T), USN
F. R. Strayer, Ensign (T), USN
M. R. Thompson, Ensign (T), USN
J. W. Young, Ensign, USNRF
R. A. McDonnell, Lieut. (jg) (PC) USNRF
D. C. Moore, Pharmacist, USN

Officers Serving in Army Transports

UNDER VICE ADMIRAL CLEAVES, U. S. N.

CONVOY OPERATIONS

B. B. Bierer, Commander USN
P. N. Olmsted, Commander, USN
S. V. Graham, Commander, USN
W. R. Sexton, Commander, USN
J. R. Defrees, Commander, USN
C. Shackford, Commander, USN
W. N. Jeffers, Commander, USN
H. W. Osterhaus, Lt. Commander, USN
L. Coxe, Lieut. Commander, USN
D. T. Ghent, Lieut. Commander, USN
G. P. Chase, Lieut. Commander, USN
A. C. Pickens, Lieut. Commander, USN
P. E. Dampman, Lt. Commander, USN
S. L. H. Hazard, Lt. Commander, USN
O. Hill, Lieut. Commander, USN
C. P. Parker, Lt. Commander, USNRF
W. L. Beck, Lieutenant, USN
B. R. Ware, Jr., Lieutenant, USN
J. L. Kauiffman, Lieutenant, USN
S. Cochran, Lieutenant, USN
R. King, Lieutenant, USN
R. M. Comfort, Lieutenant, USN
C. M. Smith, Lieutenant, NNV
F. A. Braisted, Lieutenant, USN
W. J. Gessner, Lieutenant, USNRF
W. B. Kimball, Lieutenant, USNRF
O. C. Greene, Lieutenant (jg), USN
J. B. Oldendorf, Lieutenant (jg), USN
A. Y. Lanphier, Lieutenant (jg), USN

R. C. Smith, Jr., Lieutenant (jg), USN
J. D. Smith, Lieutenant (jg), USN
C. J. Bright, Lieutenant (jg), USN
J. J. Drury, Lieutenant (jg), USNRF
R. E. Dennett, Lieutenant (jg), USN
R. S. Parr, Lieutenant (jg), USN
F. C. Lane, Lieutenant (jg), USNRF
R. D. Tisdale, Ensign, USN
P. Marshall, Ensign, USN
L. P. Safford, Ensign, USN
R. A. Awtrey, Ensign, USN
C. A. Krez, Ensign, USN
H. M. Home, Ensign, USN
A. B. Root, Ensign, USN
T. D. Warner, Ensign, USN
W. E. Miller, Ensign, USN
W. J. Forrestel, Ensign, USN
W. S. Carrington, Ensign, USN
H. J. Redfield, Ensign, USN
H. J. Grassie, Ensign, USN
K. McGinnis, Ensign, USN

OFFICERS SERVING IN ARMY TRANSPORTS UNDER VICE ADMIRAL CLEAVES

ANTILLES

Daniel T. Ghent, Commander, USN

CALAMORES

A. Staton, Lieut. Commander, USN
R. Wainwright, Lt. Commander, USN
J. C. Tyler, Lieut._ (jg), USN
C. L. Jacobsen, Lieut, (jg), USN

EDWARD LUCKENBACH

A. C. Pickens, Commander, USN
R. C. Smith, Jr., Lieut. Commander,
USN A. W. Sears, Lieut. Commander, USN
R. E. Dennett, Lieutenant, USN
W. W. Bowes, Jr., Lieutenant, USNRF
N. B. Rhoades, P. A. Paymaster, USN

EL OCCIDENTE

H. W. Osterhaus, Captain, USN

FINLAND

S. V. Graham, Commander, USN
W. J. Giles, Lieut. Commander, USN
W. J. Forrestel, Lieutenant (jg), USN
R. F. Skylstead, Lieutenant (jg), USN

LENAPE

P. E. Dampman, Commander, USN
Chauncey Shackford, Commander, USN

LOUISVILLE

J. P. Jackson, Commander, USN

MALLORY

G. P. Chase, Lieut. Commander, USN
R. L. Love, Lieutenant, NNV
R. E. Keating, Lieutenant (jg), USN
A. G. Burt, Lieutenant (jg), USNRF
M. J. Dierlam, Lieutenant (jg), USNRF

MONTANAN

P. N. Ohnsted, Captain, USN

MOMUS

W. N. Jeffers, Captain, USN

PASTORES

O. Hill, Lieut. Commander, USN
R. S. Parr, Lieutenant, USN
R. A. Awtrey, Lieutenant, USN
M. Comstock, Lieutenant (jg), USN